ELECTRICAL DISTRIBUTION IN BUILDINGS

C. Dennis Poole
CEng, MIEE

SECOND EDITION

Revised by

Trevor E. Marks
ARTCS, CEng, MIEE, MInstMC, FInstD

Consulting Editor:
E.A. Reeves

**Blackwell
Science**

Second edition © 1994 The Estate of
C. Dennis Poole, and T. E. Marks
First edition © 1987 The Estate of
C. Dennis Poole

Blackwell Science Ltd
Editorial Offices:
Osney Mead, Oxford OX2 0EL
25 John Street, London WC1N 2BL
23 Ainslie Place, Edinburgh EH3 6AJ
350 Main Street, Malden
    MA 02148 5018, USA
54 University Street, Carlton
    Victoria 3053, Australia
10, rue Casimir Delavigne
    75006 Paris, France

Other Editorial Offices:
Blackwell Wissenschafts-Verlag GmbH
Kurfürstendamm 57
10707 Berlin, Germany

Blackwell Science KK
MG Kodenmacho Building
7-10 Kodenmacho Nihombashi
Chuo-ku, Tokyo 104, Japan

The right of the Authors to be identified as the
Authors of this Work has been asserted in
accordance with the Copyright, Designs and
Patents Act 1988.

First Edition published 1987
Reprinted 1989
Second Edition published 1994
Reprinted 1995, 1998, 1999, 2000

Set by Setrite Typesetters, Hong Kong
Printed and bound in the United Kingdom
at the University Press, Cambridge

The Blackwell Science logo is a
trade mark of Blackwell Science Ltd,
registered at the United Kingdom
Trade Marks Registry

DISTRIBUTORS

Marston Book Services Ltd
PO Box 269
Abingdon
Oxon OX14 4YN
(*Orders*: Tel: 01235 465500
        Fax: 01235 465555)

USA
Blackwell Science, Inc.
Commerce Place
350 Main Street
Malden, MA 02148 5018
(*Orders*: Tel: 800 759 6102
        781 388 8250
        Fax: 781 388 8255)

Canada
Login Brothers Book Company
324 Saulteaux Crescent
Winnipeg, Manitoba R3J 3T2
(*Orders*: Tel: 204 837 2987
        Fax: 204 837 3116)

Australia
Blackwell Science Pty Ltd
54 University Street
Carlton, Victoria 3053
(*Orders*: Tel: 03 9347 0300
        Fax: 03 9347 5001)

A catalogue record for this book is
available from the British Library

ISBN 0-632-03256-1

Library of Congress
Cataloging-in-Publication Data

Poole, C. Dennis.
  Electrical distribution in buildings/
C. Dennis Poole. – Rev.
2nd ed./Trevor E. Marks.
      p. cm.
  Includes bibliographical references and
index.
  ISBN 0-632-03256-1
  1. Electric power distribution – Great
Britain. 2. Buildings –
Great Britain – Electric equipment.
I. Marks, Trevor E.
II. Title.
TK3091.P66 1993
621.319′24 – dc20   93-24612
                                      CIP

For further information on
Blackwell Science, visit our website:
www.blackwell-science.com

Contents

Preface

Although changes take place and are absorbed in the daily routine of business, it is not until one starts to revise a book that one realizes just how much change can take place in a short space of time.

Many changes have occurred since the book was first published in 1987, especially with statutory regulations. The Electricity Supply Regulations of 1937 were replaced with the Electricity Supply Regulations 1988, these in turn were amended by The Electricity Supply (Amendment) Regulations 1990 to take into account the electricity supply industry being privatized. In 1989 two more statutory regulations were introduced, The Low Voltage Electrical Equipment (Safety) Regulations 1989 and The Electricity at Work Regulations 1989, the latter having a profound impact on all persons who use electricity at work. In 1991 we also had the Smoke Detectors Act concerning the installation of smoke detectors in new domestic properties. These were in addition to a host of other legislation.

The beginning of 1993 saw the implementation of the 16th Edition of the IEE Wiring Regulations and it being made into a British Standard to enable British installation rules to be considered by CENELEC. In January 1993 the title of the IEE regulations became 'BS 7671: 1992 Requirements for electrical installations (The IEE Wiring Regulations).' However, in this book the original title 'IEE Wiring Regulations' has been retained.

In addition to the changes brought about by legislation and the 16th Edition of the IEE Wiring Regulations, changes also occurred in British Standards. The British Standards for HRC fuses were changed, as were the standards for emergency lighting and protection of structures against lightning to name but a few.

All this has meant a considerable amount of rewriting has had to be carried out to bring the book up-to-date, but I was pleased that I was asked to revise the book since I knew Dennis Poole and in particular John Vollborth, who I became friendly with whilst I was lecturing on the wiring regulations for the IEE.

The philosophy behind the original book has not changed and the original chapter titles have been retained.

Trevor E. Marks

Acknowledgements

The following illustrations are reproduced with the permission of the companies listed below:

Fig. 1.6 — Courtesy of ABB-Wylex Sales Ltd; Fig. 2.5 — Courtesy of Petbow Ltd; Fig. 6.3 — Courtesy of Hugh King, Thorn EMI Lighting; Fig. 6.4 — Courtesy of Hugh King, Thorn EMI Lighting; Fig. 6.6 — Courtesy of Olson Electronics Ltd, London; Fig. 9.3 — Courtesy of Merlin Gerin; Fig. 9.4 — Courtesy of Merlin Gerin; Fig. 11.6 — Courtesy of MK Electric Ltd; Fig. 11.8 — Courtesy of MK Electric Ltd; Fig. 12.2 — Courtesy of Power Breaker Ltd; Fig. 12.3 — Courtesy of Power Breaker Ltd; Fig. 13.1 — Courtesy of Hugh King, Thorn EMI Lighting; Fig. 14.1 — Courtesy of Trend Control Ltd; Fig. 15.2 — Courtesy of Chloride Bardic Ltd; Figs 15.5, 15.6 and 15.7 — Courtesy of Chloride Bardic Ltd; Fig. 17.6 — Courtesy of W.J. Furse & Co Ltd; Fig. 17.7 — Courtesy of W.J. Furse & Co Ltd; Fig. 18.2 — Courtesy of Fireproof Engineering Ltd; Fig. 19.2 — Courtesy of Avo International; Fig. 19.4 — Courtesy of Avo International; Fig. 19.6 — Courtesy of Avo International and Fig. 19.7 — Courtesy of Avo International.

Chapter 1
Electricity Supplies

Unless an installation is provided with a suitable and adequate supply, whether from private generation or a regional electricity company, an impossible situation is almost certain to arise as the consumer may find that he is limited with regard to the maximum loads that he can connect or the type of equipment that he can use, particularly with commercial and industrial installations. In domestic premises, unless they are situated in extremely isolated locations, it is unusual for there to be problems. Exceptions would be where the regional electricity company has no adjacent distribution system and has to install, for example, an 11 kV light-line extension and pole-mounted transformer equipment.

In all cases, however, it is essential to apply to the appropriate electricity supplier in good time.

Agreement with regional electricity company

As the majority of electrical installations require a supply from an electricity supplier which, in the UK, means one or other of the regional electricity companies, one of the most important – but too often one of the last to be considered – requirements is the negotiation of a suitable service and the most advantageous tariff for the consumer with the appropriate supplier.

Although the British Isles has probably one of the most effective supply networks in the world, experienced engineers are aware that there may be limitations on the additional loads imposed by new buildings and extensions on existing installations. However, it is too frequently taken for granted that, somehow or other, the regional electricity company will solve any problem if or when it arises, without difficulty. It should be realized that, at the least, any work necessitated by the supplier may impose unanticipated costs while, in very rare cases, it may prove completely impossible to provide a suitable supply at all. This is even more probable in certain areas of the world, particularly where the supply is weak or the load too much for the network. It is, therefore, essential that an approach be made to the electricity supplier at the conceptual stage of a completely new project and, at the latest, before connection of any

1

additional loads. In the latter case, it should be remembered that, in the UK at least, the regional electricity company may disconnect any installation completely if the consumer's installation is not constructed, installed, protected and used so as to prevent danger and not to cause interference with the supplier's system or the supply to other consumers.

Additionally, in order to comply with the 16th Edition of the IEE Wiring Regulations it is not possible to commence the design of a new installation until both the prospective fault level at the origin of supply and the external impedance have been obtained, both being factors which only the regional electricity company can provide, before the supply is available.

Prior to entering into discussion with a regional electricity company's representative, it is of benefit to have assessed the probable maximum demand, load factor (i.e., average load divided by peak load) and consumption of the proposed installation. However, when dealing with a project of any size, it is usual to find that there will be both commercial and mains development engineers involved who have a very wide experience and are prepared to offer their knowledge to a prospective consumer.

The UK electricity supply industry comprises some fourteen regional, individual, private companies who distribute most of the electricity to the consumer. Although each of these private companies can and do generate electricity the bulk generation of electricity is carried out by Power Gen, National Power, Nuclear Electric and Scottish Generators. The electricity generated being distributed into each region by the National Grid Company, which is owned by the regional electricity companies.

Figure 1.1 shows the geographical areas of the UK regional electricity companies.

Tariffs

Large consumers of electricity can negotiate for their own individual supply of electricity direct with one of the generating companies, who are responsible for bulk generation, or with one of the regional electricity companies.

The regional electricity companies are not confined to their own geographical area and can trade anywhere. For instance, NORWEB plc can trade in the area covered by East Midlands Electricity, they can even supply customers in the East Midlands with electricity, although for the moment, the customer must have a certain maximum demand before he can start to negotiate with the suppliers of electricity.

Fig. 1.1 UK regional electricity companies.

As far as tariffs are concerned the general pattern between the regional electricity companies presents three options to the prospective commercial and industrial consumer, as follows:

(1) General purpose tariff.
(2) Low voltage maximum demand.
(3) High voltage maximum demand.

The above may not always be the case, especially since privatization of the regional electricity companies, whose commercial engineer should be consulted.

General purpose tariff

This is for the relatively small installation and consists of a quarterly standing charge plus a unit charge (applicable to lighting, heating and power) which may be on a reducing scale as the number of units consumed increases. An additional possibility that may be available is a reduction in the charge for units metered during a night-time and/or weekend period.

Low voltage maximum demand tariff

This may consist of several component charges covering, for example, a monthly charge, availability, actual maximum demand, units consumed and a fuel adjustment. While the latter is related to Platt's oilgram price of 1% sulphur heavy fuel oil and, therefore, is not controllable by the regional electricity companies, the other components may also be subjected to variations governed by the suppliers' commercial requirements and are clarified as follows.

Availability

The maximum demand agreed between the consumer and the electricity supplier becomes the authorized supply capacity and for supplies up to 100 kVA the supply is usually agreed in blocks of 5 kVA. Supplies larger than 100 kVA are agreed in larger blocks.

The availability charge is made on the agreed authorized supply capacity unless this value is exceeded, in which case it is made on the higher value for the next 12 months. If the higher value of maximum demand is exceeded before the 12 months has expired the charge is then made on the new higher demand and the 12-month period then starts again. For example, if the authorized supply capacity was 50 kVA then the availability charge would be made on 50 kVA each month, if the maximum demand (which is the authorized supply capacity) is increased to 53 kVA just for the month of March, then the charge each month would be made on 55 kVA until April the following year, if, however, the demand went up to 58 kVA just for the month of June then the charge each month would be made on 60 kVA until July the following year. If after June the demand fell back and did not exceed the original 50 kVa for the remaining months, then in July the following year the charge would fall back to that levied on 50 kVA.

It can be seen that the consumer who occasionally exceeds his authorized supply capacity is penalized for the following 12 months. If the authorized supply capacity was exceeded every month then the supplier would negotiate a new contract, which may involve the consumer in a capital contribution if the excess demand cannot be permanently met from the existing local network.

Maximum demand

The charge is made for the maximum demand per month for each kVA of monthly maximum demand as follows:

March to October	Nil
December and January	£x per kVA
November and February	£y per kVA

The charges are highest in the months when electricity is in peak demand. Increased demand during the winter months would require capital costs of providing the capacity which would then be unused during the remainder of the year. The actual months and charges can vary from region to region.

Modifications are also available to these tariffs for those consumers whose highest demands occur outside the hours of 7.00 AM to 8.0 PM on weekdays or for those consumers whose highest demands occur between midnight and 8.00 AM each day.

Unit charge

This is related to the actual kWh metered, with units metered during a night-time period being considerably lower in price than daytime units. The lower price may be of the order of 40 to 50% of the higher.

High voltage maximum demand tariff

This tariff is worked out in the same way as the low voltage maximum demand tariff with the difference, of course, that the metering is effected on the high voltage side of the supply. Because of this, any costs incurred by the consumer in the transformation of the supply voltage are also metered and charged. Taking the supply at high voltage usually results in slight reductions in the various component charges relative to the low voltage maximum demand tariff, but it is wise to ensure that low-loss transformers are utilized where available.

Additionally, should this tariff be applicable, the consumer is responsible for the purchase and installation of the switchgear and transformers required, equipment which entails the employment of staff who have had adequate training and experience in the operation and maintenance of this type of equipment. Alternatively, a maintenance contract with the regional electricity company can be negotiated.

Metering

There is no obligation on an electricity supplier to provide any other metering than that required to obtain the basic data to enable tariff charges to be applied. While this may be adequate for the smaller installations, it does not give sufficient information to allow a larger consumer to allocate costs to his various facilities or to control consumption. Consequently, it may be advisable, perhaps even essential, for the prospective consumer to take such possible requirements into consideration in the planning stages and to have the necessary metering equipment built into his switchboards. Although this will increase the initial cost of the switchgear, it will prove more economical than having to add metering at a later date.

Operating cost control

Maximum demand limitation

Although electricity charges on flat-rate tariffs (domestic) are not affected by maximum demand penalties, this is not the case with maximum demand tariffs and, as explained earlier the resulting charge may constitute a large proportion of each electricity account. Consequently, it is to the benefit of the consumer to ensure that the maximum demand is kept to the lowest possible level.

In many cases the highest maximum demands are recorded during start-up periods − first thing in the morning, after the midday break or when canteen staff commence operations − and, therefore, it is to the consumer's advantage to ensure, as far as possible, that heavy machinery, or machinery which demands a high starting peak, is adequately monitored so that sequence start-up can be adopted. Where large groups of machines are supplied from one source this may be accomplished easily, and relatively economically, by automatic means built into switchboards or motor control centres but, alternatively, much success can be obtained by adequate training and good housekeeping.

Power factor correction

Maximum demand in commercial and industrial premises is generally measured in kVA and, therefore, the effects of a low power factor may be considerable. Not only will this affect the maximum demand charge but also, due to the higher than necessary currents, it will increase running costs as the I^2R (copper) losses will be greater than necessary. It is advisable, therefore, for consideration to be given to the installation of power factor correction equipment. This is not applicable to domestic installations as the tariffs do not include power factor clauses.

While, on very large factory installations, correction may be provided by means of over-excited synchronous motors, the majority of consumers should confine their requirements to static equipment with, possibly, automatic control for switching capacitors in or out in accordance with the situation or, alternatively, by installing correction equipment at the point of usage, e.g. on motors. This latter is probably the most economical method from the point of view of capital cost and, as a capacitor will only correct power factor on the supply side, will have the maximum effect on the whole of an installation. The disadvantage is that correction only applies to the drive to which it is connected and, therefore, has to be repeated on similar machines. As the life of such equipment is of the order of 25 years, this should not present any serious maintenance problems. For safety capacitors must contain discharge resistors.

Examples of tariff benefits obtained from power factor correction are published in numerous textbooks, in capacitor manufacturers' catalogues and similar literature. However, as installations and usage vary so much, the consumer would be well advised to seek help from supply specialists.

While, logically, the greater benefits are to be obtained from a power factor of unity, it should be noted that the optimum for practical purposes is approximately 0.97. Below this level, the capacitive correction required is virtually proportional to the improvement gained while, above it, the capital cost is unlikely to be recovered within an acceptably economical period of time.

A diagram such as Fig. 1.2, if drawn to scale, is one method used to calculate the amount of power factor correction required to provide improvement. AB represents kW, AD is kVA at a power factor of 0.85, AC is kVA at 0.97 power factor and BCD the kVAr. Therefore CD gives the reactive capacity (kVAr) to correct from 0.85 to 0.97 and BD from 0.85 to unity. The geometrical angles of lag are also shown on the diagram.

The reader will be familiar with this and also with the mathematical

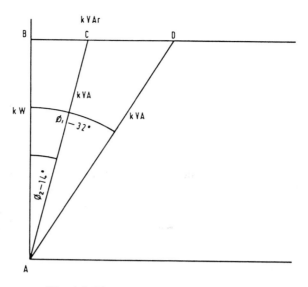

Fig. 1.2 Phase angles and power factor.

method; however, an example of the latter follows, assuming a true power of 100 kW with the above power factors.

kVA = AD = 100/0.85 = 117.6: AC = 100/0.97 = 103.1 kVA

As BD = tan∠BAD × AB, from trigonometrical tables obtain the tangents of the angles represented by cos 0.85 and 0.97, i.e. 0.63 and 0.25. Therefore:

BD = 100 × 0.63 kVAr = 63 kVAr
BC = 100 × 0.25 kVAr = 25 kVAr

From the above, 63 − 25 = 38 kVAr will be required to correct from 0.85 to 0.97 and a further 25 kVAr to bring the power factor to unity.

As power factor correction capacitors are manufactured in standard ratings, which are banked to provide the required capacity, and the combinations are not infinitely variable, it is impractical to use mathematically precise quantities for the calculations and, therefore, one or two decimal places are adequate for a working situation.

As already mentioned, a low power factor causes unnecessarily high currents for a given true power which increases copper losses and may, in fact, require larger cables to be installed than would normally be used. However, the different regional electricity companies are not consistent in their choice of preferred power factor at which penalties are imposed, and improvement above that is then only viable when the cost is recoverable against the above losses etc. As the preferred figure ranges from 0.85 for the South Eastern region to 0.98 for several others, the amount of power factor correction equipment, and therefore the cost involved, also varies considerably according to the area in which the designer is concerned.

Building management systems

During recent years there has been increasing awareness of the necessity to avoid wasting energy which has led to the growth of energy management systems in buildings. As with power factor correction, inexpert application may lead to a great deal of expenditure that will only be recovered over an extremely long period of time due to the relatively high cost of many of the systems available. However, specialists are available to advise on the best application for such systems, and where the need exists it is obviously more economical to install a management system during construction than at a later stage.

For the smaller consumer, a great deal may be accomplished without incurring too great an increase in capital expenditure. Manufacturers are producing energy-saving equipment for both lighting and power requirements, while installation designers are giving more thought to economical switching systems, particularly for lighting circuits, rather than the mass switching methods that were so popular some years ago. Regarding the latter, it is unfortunate that many of the lighting design handbooks available appear to emphasize the importance of average illumination levels and pay small regard to the fact that all illumination is required to carry out specific tasks and, therefore, that it is at the working area where adequate lighting is essential. In many industries and commercial undertakings, high levels of illumination are only required in certain areas, with much lower levels being quite adequate for access purposes. Very often, the greatest economies are related to heating, ventilating and air conditioning systems which should be incorporated in any building management system for maximum effect.

Without doubt, the most cost-effective method of energy management, although not the easiest to apply, is the correct training of employees to encourage them to switch off local equipment which is not required, e.g. lighting, fans, manually controlled heaters and machines, etc. but, obviously, they must be provided with the facilities to enable them to do so. Further consideration of the subject of energy management is included in Chapter 14. As far as lighting is concerned for areas that are not always fully occupied the passive infrared (PIR) detector is a useful means of switching off the lighting.

Fault level implications

The matter of fault levels was referred to at the beginning of this chapter with regard to the origin of the installation, but it is of equal importance wherever protective equipment is installed throughout an installation. This is emphasized to a far greater extent than ever before in the IEE Wiring Regulations particularly in the case of earth fault protection.

It must be realized that, if a protective device is only rated for running and overload protection, it may not be adequate for fault protection (except in the case of a high resistance fault). The result could be the complete disintegration of the device and, possibly, injury to anyone nearby. It is essential, therefore, to calculate the prospective fault current at the location of every protective device and ensure that it is capable of operating satisfactorily in such an event. Should this not be possible, as in the case of some motor starters, then consideration must be given to adequate back-up protection immediately behind such equipment.

Further possibilities that may arise in the case of inadequate protection are that, in the case of contactors or circuit-breakers, contacts may weld together, allowing a fault to persist and lead to the complete destruction of circuits and equipment supplied by the device or, alternatively, for the fault to be undetected and create a rise in potential of adjacent, exposed or extraneous conductive parts. This could introduce dangerous situations for personnel through indirect contact or by starting a fire. In any building to which the IEE Wiring Regulations apply, whether domestic, commercial or industrial, a protective device must disconnect a fault within the equipotential zone on fixed equipment within 5 s and on socket-outlet circuits for portable equipment within 0.4 s, when the voltage to earth is between 220 V and 227 V. The exception is, when Table 41C of the IEE Wiring Regulations is used where irrespective of the value of the voltage to earth disconnection can be 5 s even for socket-outlet circuits or hand-held equipment.

However, these limitations apply to relatively high fault currents, measured in amperes, whereas equipment specifically designed for earth fault protection (residual current devices) will operate much more quickly and by currents measured in milliamps. Consequently, the use of such devices provides a far greater degree of safety and should certainly be used in addition to other protective devices in areas which are likely to contribute to greater hazards, e.g. concrete floors, highly humid atmospheres and where there is the possibility of mechanical damage. It is important to appreciate that the greater the sensitivity of residual current devices (RCDs) installed, i.e. the lower the milliamp rating, the greater the degree of safety that should be obtained. However, electrical plant and equipment may have an inherent leakage which, although of no consequence on an individual unit, may build up to such a level through multiple installations, such as banks of fluorescent luminaires, that the rating of an RCD will be exceeded, causing it to trip. This was experienced, in fact, on a large hospital installation and was rectified by replacing the 10 mA RCDs installed with 100 mA rated units, which only gave protection against fire.

Permitted maximum leakage currents may be found in the applicable British Standards or obtained from manufacturers. In the case of luminaires, the standard is BS 4533.

High voltage intakes

As indicated in the reference to high voltage maximum demand tariffs, the necessity of utilizing a high voltage supply must entail higher initial capital costs, the employment of qualified personnel (on either a temporary or a permanent basis) and the running, copper loss and maintenance costs incurred by transformers, all of which may present problems for the consumer. However, the necessity for a high voltage intake is usually dictated by the regional electricity company's policy with regard to load, the dividing line generally being between the 250 to 500 kVA levels.

Until fairly recently, consumers were encouraged to utilize high voltage switchboards which were combined with the regional electricity company's and, therefore, choice was limited to the type favoured by each particular company. Mainly for operational reasons, this system, which obviously presented a very neat appearance, has been abandoned in favour of separate switchgear with the two switchboards connected by cable. The consumer is responsible, of course, for the latter, although the regional electricity company effects the jointing to its equipment. Separate equipments also eliminate the problems inherent in dual authority. While, unfortunately, this usually means that the consumer has to provide more space and, perhaps, separate lockable switchrooms, all of which entail extra costs, it also means that the choice of switchgear is far wider than under the original system and may range from a simple oil-filled fuse-switch up to the more refined, but also more expensive, types such as SF_6 or vacuum breakers (Chapter 9). Cost, therefore, may be one of the deciding factors.

As with any load throughout an installation, the protective switchgear must be adequate for the duty which, in this case, is the maximum rating of any associated transformers. A further matter of importance is the prospective fault level on the electricity supplier's system which, in the case of high voltage is far greater than any that will be encountered on a low voltage network, of the order of 13 kA at 11 kV. However, the choice of equipment is simplified to a certain extent as the technical parameters are standardized in the appropriate British Standards and, additionally, manufacturers generally limit their production to specified ranges such as, 250 MVA.

One further disadvantage, from the viewpoint of cost if no other, is that the protective equipment requirements are usually more extensive than with a low voltage supply, entailing the use of relay systems that require voltage and current transformers. On high voltage intakes, the least that should be considered is, overcurrent, earth fault protection and fault-current protection.

Depending upon the extent and dispersal of the load in any installation, the choice of systems lies between a radial or a ring system or a combination of both (Chapter 4). Where the total load can be carried by just one transformer an installation is relatively simple, consisting of the transformer, the consumer's switchgear and the regional supply company's intake; in this case, the whole should preferably be located in one area as near as possible to the main load centre. The only difficulty of any consequence here is that electricity suppliers are rarely enthusiastic about long lengths of their high voltage cable network on consumers' premises (even though the consumer has to contribute to the cost) and, therefore, the location has to be a compromise between what is most desirable for the consumer and what is acceptable to the supply company. This does not mean that an unreasonable attitude is ever taken, but it has to be accepted that, in the event of a fault, the supply company must have easy access and this cannot always be guaranteed by a consumer.

On a more extensive site, where there are a number of load centres, there is a greater possibility of the requirement for several transformers and, therefore, a choice of system between radial, ring or combination. There are arguments for and against all types of system; in the case of a radial system, provided that the protection is adequate, a cable fault will only affect the connected load centre while with a ring cable fault some disruption may occur at other centres until the fault has been isolated, although this again may be avoided if directional inverse definite minimum time limit protection is installed. On the other hand, the ring system allows for a faulty cable to be isolated and supply to be restored or maintained to all load centres. Probably, from the point of view of security of supply, the ring system is preferable.

One benefit of a ring system may be lost by the installation of the return cables in too close a proximity to the outgoing cables for, although it may be more economical to lay them in the same trench, a fault on one could, for example, quite easily affect the other directly. There is also the increased danger of both cables being damaged by equipment such as mechanical excavators or by ground subsidence.

Space requirement, high voltage equipment

The three main factors which dictate the amount of space that the consumer must allocate for a high voltage supply are the following.

Firstly, depending upon whether supply is to be provided by a radial feeder or linked on a ring main, space has to be provided for a single circuit-breaker or for a ring main unit. While the sizes of these may vary between manufacturers, they have been standardized to a very great extent and an average per unit is of the order of 600 mm wide by 1200 mm deep with a height of 2000 mm. A complete ring main unit, therefore, is the length of three units (1800 mm).

In addition, a clearance area must be provided all round the switchgear to allow access for the initial jointing of cables and for the later operation and maintenance of the equipment. As an absolute minimum a clearance distance of 600 mm behind and at each end of a switchboard (excluding provision for future extension) and 900 mm at the front is adequate in most cases. Frontal clearance should be additional to any space required for the complete withdrawal of a circuit-breaker.

The Electricity at Work Regulations 1989 specify that adequate working space, adequate means of access and adequate means of lighting shall be provided to prevent injury.

Secondly, the consumer's high voltage switchgear which, as mentioned earlier, is now usually separate from the electricity supplier's switchgear and requires a similar space requirement to that already given, although some saving may be gained by the installation of any transformers in proximity as, obviously, only one access width is required between the equipments. In this case, however, great care must be taken with regard to the layout to give maximum safety to any personnel carrying out switching or maintenance in the relatively unlikely event of a transformer explosion.

Finally, as modern types of transformers, particularly air or nonoil insulated, require little in the way of maintenance, a clearance distance of

600 mm on each side is quite acceptable provided that the area is adequately ventilated. However, it is important that provision for containment of the content of oil-filled transformers (Fig. 1.3) must be made by means of a surrounding trench or suitable drainage to an adjacent pit although if the site is near a river the National Rivers Authority should be consulted (Fig. 1.4). Most, if not all, of the regional electricity companies have standard layout drawings available which they will provide to the prospective consumer and, although they may tend to be on the large side, the installer must make such provision for their equipment and, at least, be guided by them for his own. The regional electricity companies' drawings do not usually allow for any future additions. Consequently, consideration should be given to what demands might be required in the future, before proceeding with a new building.

Low voltage intakes

The choice of system is effectively limited by the availability from the regional electricity companies which have standardized on voltage and frequency to 415 V three-phase or 240 V single-phase at 50 Hz. The only decision to be made by the consumer is, therefore, whether he requires single- or three-phase, a decision which, in any case, could be taken out of his jurisdiction if his anticipated load exceeds the supplier's capability in the case of single-phase. Further deciding factors, besides the above, are: firstly, security, a complete loss of power is less likely with a polyphase supply; and secondly, the characteristics of the equipment to be installed, the larger ratings of rotary equipment are invariably three-phase, e.g. air conditioning and freezer plant. Where lower voltages are required (or direct current), the consumer has to make his own provision by step-down transformers, batteries or rectifiers.

With regard to equipment, the consumer has a wide choice in both type and manufacturer. The regional electricity companies always provide a service intake and the essential metering into which will be connected the consumer's meter tails. However, unlike high voltage supplies, it is still possible to incorporate these in a combined switchboard. Between the metering and the consumer's main distribution board there must be some means of isolating the complete installation. The type of equipment required is a switch capable of switching the full load of the installation.

Space requirement, low voltage equipment

It is not possible to be specific regarding the amount of space required at

Fig. 1.3 Typical 1000 kVA oil-filled distribution transformer with oil conservator.

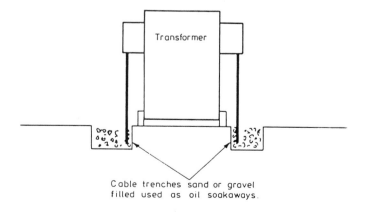

Fig. 1.4 Combined cable trench/oil retention system.

Fig. 1.5 Dorman Smith 'Loadframe' low voltage switchboard suitable for use as a mains supply intake or site distribution.

the intake as many factors have to be taken into consideration. Apart from the basic requirement that any equipment installed must be adequate in all respects for the purpose for which it is to be used, as clearly indicated in the Electricity at Work Regulations 1989, it is generally convenient to associate the main distribution board with the intake and this will, obviously, affect the total size and, therefore, the amount of space needed. Other factors governing space requirements are the choice of type of assembly, e.g., wall-mounted for a small installation, frame-mounted on a wall or free-standing, or a cubicle type for the larger installation while, with regard to equipment, a 2000 A moulded case

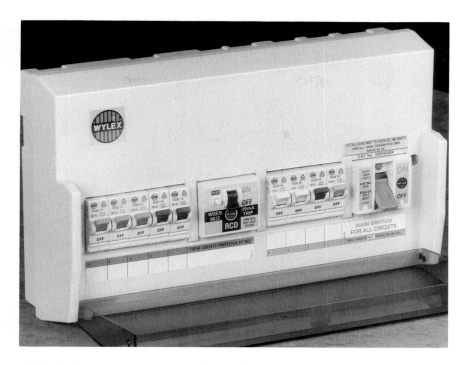

Fig. 1.6 Modern consumer unit having facilities for locking devices, timers, contactors, transformers etc. MCBs and HRC fuses are interchangeable and the unit will accept direct fitting MCB/RCD combinations. (Courtesy of ABB–Wylex Sales Ltd.)

circuit breaker may be smaller than a 500 A fuse-switch. In addition to the physical dimensions, two other matters of great importance are that:

(1) Every installation has to be protected against overcurrent by suitable devices (IEE Regulation 130–03–01).
(2) Adequate and safe means of access and working space must be afforded for operation or attention (IEE Regulation 130–07, 513–01, 529–01–02 and Regulation 15 of the Electricity at Work Regulations 1989).

Factory-built assemblies of low voltage switch and control-gear are dealt with in BS 5486. Examples of low voltage intake equipment are shown in Figs 1.5 and 1.6.

Chapter 2
Private Generation Supplies

Despite the stability of supply in Britain there are occasions, such as during heavy thunderstorms or the rare major failure at a generating station or on the grid system, when a failure of supply occurs and, although it may be restored in a relatively short time, it is rarely possible to notify consumers of the duration of these shut-downs. Such breaks may have very serious consequences for many consumers if they last for several hours as, inevitably, production is brought to a standstill whether in offices, shops or factories.

In many parts of the world, particularly in Third World countries, public supply is much less reliable than in Britain as, in some cases, the supply systems have not yet developed to the same extent due to the vast areas to be covered, the difficulties created by the terrain, economic considerations or the more extreme weather conditions. Since the middle of this century, industry and commerce have developed enormously worldwide, so that continuity of supply is more than ever necessary.

For the above reasons private generation facilities may have to be considered, in addition to the fact that the public supply may not be adequate or available at the required location. The latter is highlighted, of course, in the case of offshore oil rigs. Additionally, the nature of supply tariffs may provide reasons of economy by allowing the consumer to transfer excessive peak loads from public to private supply, thereby avoiding maximum demand impositions.

Base load systems

In the UK, before the development of our large generating stations and the 132 and 400 kV grids particularly, industries such as large steelworks, chemical plants, coal mines and those in remote areas found that the public supply was not capable of providing their requirements and so they installed their own generating plant. There are still some large users of electricity, particularly those which require passout steam for production purposes, who operate their own power stations.

In other countries, private generation is far more probable than in the

UK. There are numerous processes which require heat which can be extracted from a generating system in the form of waste-heat, low pressure steam and hot water. In such cases, it may prove economical to include private generating facilities for supply and also for use in a manufacturing process. In the steel industry, for example, coal-fired boilers have been used to provide coke (then used in blast furnaces), benzol and tar while, in much smaller industries, the cooling water for a diesel engine may be passed through a heat exchanger and used for a wide variety of heating purposes.

With a very large industrial complex, it is more probable that private generation would be used as the only source of supply because of the heavy demands, measured in megawatts; consequently, a suitable public supply may not be available. Such private generation is normally provided from a number of sets and the probability of a complete failure is extremely low.

Connection with regional electricity company

For the size of commercial and industrial installation mainly under consideration, the foregoing situation is unlikely as the capital cost of such independence is extremely high, it invariably necessitates two or more generators to obtain a firm supply and facilitate maintenance, and one still has to consider the running, operational and maintenance charges. In these cases, therefore, there is the possibility of combining private generation with the public supply and using either as the main source. If a combined system is to be considered it is extremely important to discuss the proposals with the regional electricity company before any commitments are made as, in the UK, certain rights are vested in them by legislation.

Until the British statute was amended in 1983, for example, parallel operation of public and private supplies was not often encouraged, therefore, it was advisable for installations to consist of two physically separated sections or, alternatively, provision had to be made on the main switchboard, at the consumer's expense, for changeover facilities. Because of this stipulation, either each source of supply had to be capable of dealing with the total load or switching arrangements had to be incorporated to limit loads to the maximum capability of the lowest source. A further important aspect of a dual supply of this nature was that, if the public supply was used only intermittently — as, for example, in emergency situations or for maintenance shut-downs on the private plant — the capital and running costs imposed by the regional electricity company tended to be rather high. In many cases, they can insist upon certain requirements regarding the actual installation of private generating plant, such as complete segregation of a generator earthing system from their own earthing system.

When paralleling is permissible, generator characteristics must be compatible with the suppliers' supply and, as a consumer has no control over this, control gear must be capable of adjusting the voltage, frequency and synchronizing of the generator. With many of the available smaller sets, the engine has a fixed speed which is set up during manufacture and, therefore, is not suitable for parallel operation. Further problems that may arise are that, if the relative loads are inadequately controlled, sudden fluctuations may alter the engine speed, and therefore the frequency, and, in the event of engine failure, motoring on the public supply could occur.

Difficulties created for the regional electricity companies by parallel operation are that if their system fails, it may be inadvertently energized from private generation, with fatal consequences for repair teams; circuit protection is more complicated because, through the high voltage and low voltage networks, there could be the possibility of paralleling generators of the order of 250 to 500 kVA with a major substation or, at the extreme, with a generating station; obviously, both supply and generating protective equipment is designed to give protection upstream on their networks and, consequently, is not sufficiently sensitive to provide any assistance to a private supplier. Further complications arise with metering at the consumer's premises as, with the possibility of parallel operations, this must be suitable for both exporting and importing energy.

All fluctuations in load to some extent affect the speed of a prime mover driving the associated alternator, the lower the mass of the set, the greater the effect. Consequently, the relatively small diesel/alternators are more susceptible to surges, however caused, than generating stations. Further, the turning moment of turbines is absolutely regular while that of reciprocating engines is not, although accurate balancing and the use of heavy flywheels can reduce the cyclic irregularity. When private generating plant is used only for the purposes of the operator, any variations can only be caused by his own installation and, provided that it has been correctly sized, will have little effect on the normal power equipment, e.g. motors, lighting, contactors, etc. However, if the set also feeds into the public supply system, over which the operator has no control, this will not necessarily be the case and heavy, rapid changes in demand will cause the set to hunt in trying to meet the demands and, in the worst situation, to drop out of synchronism.

This possibility is even more pronounced if a reciprocating engine-driven set is in parallel with steam turbines, due to the difference in cyclic regularity, and it may lead to heavy interchange currents between the alternators and possible destruction of the smaller machine. Consequently, it is necessary to discuss requirements in depth with the manufacturer of the private set before purchase if it is required to serve such a purpose.

Combined heat and power systems

Reference has already been made to the use of waste-heat from prime movers which, in the case of smaller sets, can be some 50% of the energy produced under full electrical load conditions and does not fall to any great extent when full load is reduced; consequently, considerable economy can be effected by using the heat sensibly. The use of heat exchangers on marine engines is a recognized technique, the fresh cooling water being pumped through one set of coils and the heat discharged into seawater circulating through a secondary set. This principle has only been adopted on the smaller stationary engines used on generating sets since the early 1970s, the period when all fuel costs started to escalate. As the cost of the additional equipment is low in comparison with the cost of even the smaller sets, advantage may be obtained by the incorporation of heat exchangers for the direct heating of low temperature water, e.g. for washing, showers, etc. − usually referred to as domestic facilities − or for the initial heating of boiler feed water. It is also possible, by enclosure of sets and forced ventilation through the chamber, to make use of heat radiated from the sets provided that precautions are taken to avoid obnoxious fumes being transferred into other areas.

Combined heat and power systems (CHP) have been attempted from major generating stations but have not been developed in the UK to any great extent as yet, largely due to the fact that, generally, these are remote from premises that could benefit. It has been used for district heating schemes in some countries on a wide scale, however. At present, in Britain, CHP is in use at the Drax station in Yorkshire for the heating of greenhouses for tomato-growing, where it has proved extremely successful, and a trial scheme is in operation in Pimlico which, if successful, will no doubt be adopted elsewhere. CHP generating stations are being developed similar to the one that East Midlands Electricity plc have designed at Corby.

Peak lopping generation

In many industries the load demands are extremely variable and conditions arise in which very heavy demands occur at infrequent intervals. When maximum demand tariffs are applicable with such conditions, the maximum demand charge is completely disproportional to the average load and unit charges. To avoid excessive costs, it is possible to use a generator, in parallel with the mains supply, to take the peaks, i.e. peak lopping. An additional advantage is that the set will be available for use in an emergency.

It is not easy to quantify the economical advantages of peak lopping, but an example of the cost incurred by peak demands is given below.

Monthly charge
For supplies metered at high voltage   £131.00
Supply availability
Charge per month for each kVA of chargeable supply
    capacity   £1.04
Maximum demand charge (MD)
For each kVA of monthly maximum demand November
    and February   £2.68
For each kVA of monthly maximum demand December
    and January   £7.35
Unit charges
For each unit supplied (kWh) during on-peak periods   5.30p

If the normal working maximum demand is 400 kVA with a half-hour peak demand of 500 kVA each morning when the factory starts-up, what is the additional cost incurred for the peak maximum demand?

Availability charge per month for 500 kVA = £1.04 × 500 kVA = £520.00
Availability charge per month for 400 kVA = £1.04 × 400 kVA = £416.00

Additional cost per month for 12 months   £104.00

Maximum demand charge for November and February
    500 kVA = 500 × £2.68   £1340.00
Maximum demand charge for November and February
    400 kVA = 400 × £2.68   £1072.00

                               Additional cost each month   £268.00

Maximum demand charge for December and January
500 kVA = 500 × £7.35   £3675.00
Maximum demand charge for December and January
400 kVA = 400 × £7.35   £2940.00

$\overline{}$

Additional cost each month   £735.00

$\overline{}$

Unit cost per year for the peak load only, assuming 4 weeks shut down:

$$48 \text{ weeks} \times 5 \text{ days} \times \tfrac{1}{2}\text{h} \times 100\,\text{kW} \times \frac{5.30\text{p}}{100} = \qquad £636.00$$

Total annual cost of peak maximum demand during factory start-up

Availability cost × 12   £1248.00
Maximum demand charge Nov/Feb   £536.00
Maximum demand charge Dec/Jan   £1470.00
Unit cost for peak period   £636.00

$\overline{}$

£3890.00

$\overline{}$

This annual cost, due to a half-hour peak in demand each morning, could mean that the cost of installing a 100 kVA generator would be recovered in about 10 years. However, the cost of the generator is not the only cost that has to be taken into account, the other costs that must also be considered are:

(1) Additional switchgear including synchronization equipment.
(2) Fuel storage.
(3) Running cost of the generator: labour, fuel, maintenance etc.
(4) Maintenance of associated equipment.
(5) Additional labour costs of competent personnel to operate the equipment.

Having considered all the above there still remains the problem of having sufficient time to shut the generator down for overhaul. The danger with this type of arrangement is a failure of the generating equipment, should this arise the extra 100 kW start-up power would have to be supplied from the electricity supplier negating any saving in availability charges for 12 months.

It is strongly recommended that any proposal is fully discussed with generator manufacturers and the regional electricity company, who have a great deal of experience in these matters, before any decision is made since there are many alternatives to installing a generator.

The authorized supply capacity is agreed in blocks of kVA. For demands up to 100 kVA the authorized supply is usually in blocks of 5 kVA and for 100 to 250 kVA in blocks of 10 kVA. The larger the demand the larger the multiple of kVA that is used.

When the maximum demand (chargeable supply capacity) exceeds the agreed amount, the availability charge is made to the next highest block for the next 12 months even if the excess load was a once only occurrence. If this new higher maximum demand is then exceeded during the 12-month period then the charge is made to the next higher block of kVA and the 12-month period starts again from the occurrence of the new peak demand. The supply availability charge is not reduced back to the original value until the maximum demand has not exceeded the original value for 12 months.

Where the authorized maximum demand (authorized supply capacity) is constantly being exceeded, the supplier will review the contract and if the increased demand cannot be met from the existing local network a capital contribution may be required from the customer.

Standby systems

The Health and Safety at Work Act and the Electricity at Work Regulations 1989 emphasize the necessity for safe working conditions at all times and, although there are many relatively economical methods of providing emergency lighting, this will only assist employees to escape from a potentially dangerous situation and not safeguard machinery and production processes which, in some cases, could lead to even more danger. It is for such possibilities that consideration should be given to the installation of standby generating equipment of sufficient power to enable plant to be safely closed down or essential services to be maintained. For these purposes, an autostart generating set is usually quite adequate as, from a cold start to full load, the smaller sets in particular require only a matter of seconds (Fig. 2.1).

It is essential that emergency sets operate only under specific conditions and, therefore, adequate control equipment must be incorporated in the switchgear to which it is connected. Further, essential and nonessential circuits must be made capable of segregation from each other, usually by feeding each from a different section of busbar and installing means between the two to effect the separation when the mains supply fails.

As the operation of emergency sets may be extremely infrequent, adequate test and maintenance procedures must be instituted − the same applies, of course, to any equipment subject to infrequent use − and test-

Fig. 2.1 Dale 50 kVA Panda generating set powered by a Perkins diesel engine and controlled by the Dale 7000 system provides standby power for a hospital.

discussed with the manufacturer as short duration light load test-runs are often detrimental to the engine, in which case it is preferable to imitate emergency conditions by applying additional load. Alternatively, it may be possible to use the standby set for supplying part of the load for several hours/days (Fig. 2.2) provided that agreement is obtained from the electricity supplier (particularly if this involves paralleling with the external supply).

Temporary site supplies

During the early construction phases of a new project it is unlikely that a regional electricity company will be able to provide a mains supply to assist builders and service installers, or certainly not without considerable extra cost, and, even then, it is usual for weatherproof accommodation to be required.

Fig. 2.2 Multigenerating set installation at the Royal Bank of Scotland controlled by the Dale 8000M system. The installation consists of three Dale mains failure sets powered by 1000 kVA MTU diesel engines and two powered by 720 kVA MTU diesel engines. Note the independent foundations.

To overcome this problem, it may be worth considering the provision of a package generating set which, due to the robust type of construction employed, may be transported to site and installed on any firm, level surface, a bed of railway sleepers is excellent for this purpose.

For protection against the weather, the superstructure of a site cabin may be used although it is possible, and preferable, to install a package set complete with weatherproof housing and, if necessary, acoustic treatment (Fig. 2.3).

As a further alternative, trailer-mounted sets are available which, of course, if taken on a public highway, must comply with the applicable traffic laws.

One matter which must not be overlooked is that, with this type of installation, provision must be made for earthing both the frame and the neutral of the unit.

For extensive construction sites, equipment is available for use as load-centres which is specifically designed for outdoor use in arduous conditions and, together with a package set, provides a complete site distribution system independent of mains supply. This is of particular advantage if a considerable amount of commissioning work on air-conditioning plant, lifts, etc. is required before final completion.

Fig. 2.3 Petbow 15 dBA Slimsound enclosure for 400 kW diesel generating set on construction site. Note the arrangement of the foundations.

Space requirements

As with switchgear, it is impossible to give more than general guidance on the space requirement as this depends on the size of set installed and whether or not bulk fuel supplies are to be available. The majority of sets suitable for commercial and small to medium factories are available complete with day-tanks of sufficient capacity to operate for up to 48 h. This is adequate for many peak lopping and emergency requirements (although bulk purchase of fuels usually provides an economic advantage) but base load generation entails continuous use which could be served from bulk supplies. A further factor affecting space is that this type of plant may be trailer mounted or supplied complete with weatherproof housing for external use.

Dimensions provided by Dale Electric of Great Britain Ltd are given in Table 2.1 and as, when installed within a building, working space is required all round, they may be taken as typical sizes for guidance on space requirements.

When generating sets are installed within buildings, and the sizes show quite clearly that this does not present undue difficulty, it is essential to ensure that adequate ventilation is available, by either natural or forced means. If the latter, there is also the possibility, as already mentioned, of utilizing the waste-heat to the benefit of the building and its occupants.

With equipment provided specifically for outdoor use, in addition to

Table 2.1 Dimensions related to kVA ratings.

kVA	Length (m)	Width (m)	Height (m)
4.5	1.12	0.6	0.76
25	1.73	0.7	1.07
50	2.2	0.72	1.07
100	2.1	0.89	1.27
250	3.1	1.02	1.14
500	4.0	1.35	2.05
1000	4.75	1.9	2.39

Fig. 2.4 Petbow heavy-duty acoustic enclosures for 600 kW generating sets complete with control room.

weatherproofing, it is also possible to obtain both acoustic housings and residential-type silencers which avoid noise disturbance to surrounding areas (Figs 2.4 and 2.5).

Installation requirements

As manufacturers are able to supply private generating plant with outputs in excess of 4000 kVA capacity as package units, the degree of factory

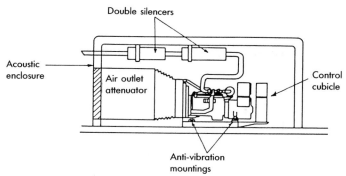

Fig. 2.5 Recommended methods for reduction of sound transmission. (Courtesy of Petbow Ltd.)

assembly is governed only by transportation and site difficulties such as limited access. In the majority of commercial and industrial installations considered here, however, it is unlikely that sets larger than 500 kVA are required and since, as will be seen from the previous section, these are of the order of 4 m in length, it is to the advantage of the purchaser to have his equipment delivered in package form.

All of the items in a package unit — engine, alternator, fuel tank and control equipment — are assembled accurately at the works on a rigid, fabricated bedplate and require only a suitable load-bearing base to accept the unit. If a set is to be installed in a basement or on a firm factory floor it is often unnecessary to provide anything further as, although a package unit must be levelled up to avoid excessive vibration, this is achieved by shimming below the bedplate. For the smaller sets, it is not even essential to bolt them down, two 85 kVA units were installed in a basement boilerhouse, with some 600 mm between them, in the mid-1970s and have given satisfactory service on many occasions without any movement whatsoever.

There are situations, however, as on the upper floors or roof of a building, where any transmission of vibration must be avoided and, in these cases, it is essential to install such equipment on antivibration pads which are also available from the manufacturers of the sets.

In the UK, all major manufacturers of this type of equipment have suitably trained staff to deal with all aspects of installation, leaving only the final cabling to the purchaser.

Chapter 3
Uninterruptible Power Supplies

The problems arising from complete breakdowns of public supply have been referred to in Chapter 2, but it is more probable that short-term interruptions will occur at more frequent intervals, many of which may be almost unnoticeable to the human eye and have no effect of consequence on power plant. Similarly, mains supply, whether from public or private sources, is subject to interference from numerous other operations which, while not causing complete failures, create fluctuations of sufficient magnitude to affect computers, microprocessors, etc.

Such interruptions and fluctuations are frequently caused by load-transfer switching operations carried out by the regional electricity companies, by the inductive effect of lightning strokes on overhead lines or by operations such as the starting and stopping of motors. The result of these, to the general public, may be no more than a slight flicker on a television screen but, in commerce and industry, they may have far-reaching effects. Particularly since the early 1970s, computers and microelectronic devices have developed from being little more than an improved high-speed form of paperwork to a highly sophisticated aid to business and to process control operations in many industries, and it is this type of equipment which is the most sensitive to variations in supply characteristics.

The extent of electricity supply systems makes them vulnerable to disturbances such as transients, surges and spikes which affect all associated consumers on the network and, where equipment is in use which is sensitive to such occurrences, it is possible to install load isolating units, in either dynamic or static form, which will ensure a 'clean' supply as long as the mains supply is available but which does not guard against complete failure of the supply. In this event, where production processes are operated which would suffer extensive or irreparable damage from a prolonged shutdown, it is essential to employ other means (Fig. 3.1).

Chapter 2 referred to generating sets for standby purposes but it was indicated that, invariably, some delay arises, even if only a matter of seconds, before such sets can take on load and, if machinery control gear includes 'no volt' protection, for example, a short break in supply causes as much disruption as an extensive failure. To overcome this problem, equipment is commercially available which ensures that a totally uninter-

Fig. 3.1 Mawdsley's load isolation motor generator sets form part of the computerized editorial system introduced in the *Birmingham Post and Mail's* headquarters at Colmore Circus.

ruptible power supply (UPS) is provided regardless of the state of mains supply. UPS also has the advantage that it eliminates the effects of the minor disturbances already mentioned such as surges, spikes, and voltage and frequency fluctuations.

There are, as with load isolation, two basic types of UPS, rotary and static, both of which have certain limitations, and these are referred to in the sections which follow.

Rotary plant

The type of generating set referred to in Chapter 2 has the disadvantage mentioned that start-up does not commence until the primary supply has failed and the control equipment operated and, while this may only be a delay of seconds, it will certainly be long enough to make reprogramming of some designs of computer necessary or even, in some cases, to cause complete loss of the program. To overcome this, UPS equipment may be installed which eliminates the delay completely, allows the user to restore to normal at his convenience and provides isolation from the mains and, therefore, protects from possible transients (Fig. 3.2).

Fig. 3.2 Mawdsley's $2 \times 40\,\text{kVA}$ single-phase flywheel type rotary UPS systems supplying power for the Network Operations Centre of Mercury Communications Ltd.

The most simple rotary UPS equipment consists of a diesel engine, alternator and automatic starter, with the addition of a flywheel and clutch.

Under normal supply conditions the alternator is driven as a motor from the incoming supply switchboard together with the flywheel. Upon supply failure, the engine is started and takes up the load, when full speed is reached, through the clutch. During the start-up period the kinetic energy stored in the flywheel maintains the alternator speed at approximately normal to maintain the supply.

A variation on this incorporates an a.c. motor between the engine and flywheel which drives the alternator; with this type the load is supplied permanently from the alternator and avoids any possible supply transients. The only effect on the installation of mains failure may be a slight drop in voltage and frequency during the transition period between failure and diesel onload.

Alternatives to the above are available which, in place of the diesel engine, incorporate a d.c. motor and a battery. The motor is permanently supplied from the incoming supply via rectifier/charger equipment but, on supply failure, the battery continues to supply the motor with, again, the flywheel maintaining the speed of the set. The duration of supply in this case is limited by the capacity of the battery and, therefore, this equipment is

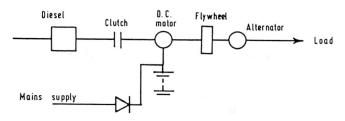

Fig. 3.3 Rotary UPS system with diesel back-up.

for short-term duty only. However, this problem is removed by the addition of a diesel engine and clutch (Fig. 3.3).

Such systems may be extended to provide parallel redundancy by installing two or more sets operating in parallel; if each set is capable of dealing with the total load, the operator has the facility to close down each set in turn to effect maintenance or to retain one or other purely as a standby system. As these installations are costly it may be necessary to use sets of smaller rating, e.g. each one adequate for half of the total load, and incorporate facilities for load-shedding in the event of failure of a set or, as a further refinement, to include bypass switching to transfer the load directly onto the mains supply (Fig. 3.4).

With all rotary equipment, static switching may be used which has a much higher response time than electromechanical devices.

A further development in UPS, although classified as rotary equipment, combines both rotary and static units. The input supply is to the rectifier/battery/inverter components with the output feeding into a rotary machine which has a double-wound stator to provide supply to the distribution.

With this type of equipment, the quality of the output from the inverter is almost immaterial as the harmonics are attenuated to a great extent by the rotary machine.

The advantages of this system are that, being in constant operation, there is no possibility of a break in supply or of voltage and frequency fluctuations and, as it is available in large ratings, it is suitable for extensive computer and similar installations. The manufacturers also claim that it provides increased reliability due to the simplified construction of the rotary machine, which has no brushes and only two bearings, and an anticipated life between failures of 10^6 hours.

Static systems

For situations requiring low power supplies from below 100 VA up to

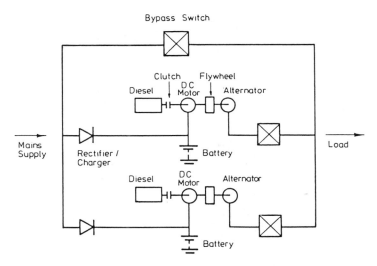

Fig. 3.4 Redundant arrangement, rotary UPS system.

5000 VA, where there are space or noise restrictions, static systems are available which, as with rotary equipment, have several variations to suit the need of the user. These small equipments are sufficiently compact for installation at point of use in offices or commercial premises where mains supply is available from adjacent existing small-power circuits.

Where the proposed loads require higher ratings, similar equipment is produced in single- or three-phase mode with capacities up to 60 kVA and 750 kVA, respectively, in single units. Space requirements are obviously greater than with the smaller units but the equipment is not unsightly and has a low noise level which allows it, if necessary, to be installed in close proximity to the end user without causing undue disruption to other activities (Fig. 3.5).

The system consists of three basic elements: rectifier/battery charger, battery and inverter. The mains supply charges the battery and the d.c. output is then inverted to a.c. to feed the sensitive load. The operations effectively filter out all mains-borne interference and ensure a stable voltage and frequency. By the use of current-chopping techniques, higher frequencies than the standard 50 or 60 Hz may be obtained which, although equally possible with rotary equipment, may be more economically achieved with static.

The basic system is designed for continuous operation but variations include an electromechanical bypass which provides automatic changeover from the mains to the battery on supply failure or a static switch bypass

Fig. 3.5 A typical static UPS power supply covering three-phase ratings from 20 to 60 kVA at 415 V, 50 Hz.

which effects a transfer within microseconds. Obviously, these changeover facilities are only possible when the output voltage and frequency are identical to the input.

The very small UPS equipment referred to is manufactured in ratings up to approximately 5000 VA and, therefore, fills the need of the smaller consumer or isolated requirements on large installations where extensive field wiring between stations may be unduly expensive or difficult. However, in order to maintain the compact size of the unit the battery is of a relatively low capacity − typically between 15 min and 2 h − and, therefore, is intended for use in continuous mode to provide a clean supply or, with the incorporation of bypass switching, as short-term standby (Fig. 3.6).

The larger ratings up to 750 kVA, providing the same facilities, are also available as package units and, although varying little in size from comparable rotary equipment, are more pleasing in appearance and avoid both the problem of noise or, with diesel engines, fuel supply and storage. Batteries for this type of equipment are generally provided separately from the charger/inverter cubicles (although they may be housed in matching cabinets), which gives the purchaser more freedom of choice

Fig. 3.6 A typical single phase static UPS system with ratings up to 5 kVA.

with his selection and location for them. The latter is of some importance as it allows separate battery rooms to be used which may be some distance away from the UPS equipment. UPS configurations are indicated in Fig. 3.7.

It will be appreciated that, even with the larger ratings of static equipment, batteries inevitably limit their operational duration in the event of a mains failure, unlike engine-driven plant which, apart from breakdowns, simply needs an adequate supply of fuel.

Redundant systems

As both rotary and static UPS systems are not infallible − particularly those which include diesel engines or are not in continuous use − and are generally associated with important services, consideration should be given to the duplication of such units. If each UPS package installed is capable of handling the total load, the operator has the facility to run them continuously in parallel sharing the load between them; should one unit fail, the remaining one takes on the full load and eliminates the possibility of a shut-down. Although the method described provides almost

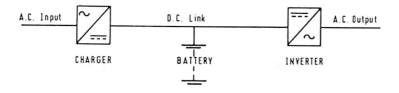

CHARGER   BATTERY   INVERTER

(a) Continuous Mode.

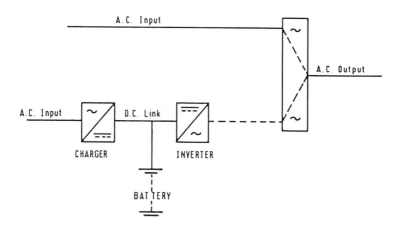

(b) Continuous Mode
Static switch bypass.

Fig. 3.7 UPS configurations.

certain protection against loss of supply, some economy may be achieved in running and maintenance costs by operating only one UPS unit at a time and accepting the possibility that the standby unit will not be available when required. With both methods there is no difficulty in carrying out maintenance in accordance with a preplanned programme (Fig. 3.8).

If economy in capital costs is required, the UPS system is designed to handle only the most essential loads and units of smaller capacity are then possible. Again, both of the above systems may be considered but single-set operation will not achieve the same level of economy and may not be worthwhile compared to the costs incurred by a complete shut down.

Unless a UPS system is intended to serve a complete installation of

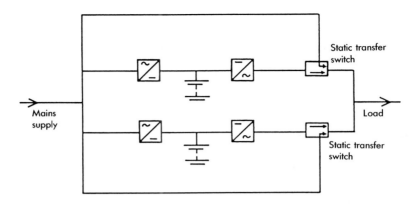

Fig. 3.8 Parallel redundant static UPS.

computers or similar equipment, extra care must be taken by the designer to ensure that other services, e.g. motor supplies, are completely screened or segregated from the UPS as, otherwise, mains interference may well be reintroduced and defeat the intention.

Chapter 4
Distribution Systems and Equipment

High voltage distribution

Although domestic and the majority of commercial installations are adequately served from low voltage distribution systems, there are many industrial and some commercial requirements for high voltage supplies with extensive distribution throughout the sites or buildings. For these, the regional electricity company provides the main supply point through its own switchgear complete with bulk metering, and the consumer is then responsible for the installation of suitable high voltage switchgear, all cabling and the required number of site substations. The consumer's main high voltage switchboard is a convenient location for feeder metering as this only needs the installation of one voltage transformer to cater for all the outgoing circuit-breakers. This metering summates the loads on the system while covering individual feeders.

In Chapter 1 it was indicated that the choices for high voltage distribution systems are either radial, ring main, or a combination of both.

Radial feeder systems

For simplicity in both installation and operation, the radial system is the most suitable but it has the disadvantage that it does not provide an alternative supply in the event of a major fault. The consumer's main high voltage switchboard incorporates a circuit-breaker for each feeder, with appropriate protection, and possible meters, from which suitable provision is required for cables to be installed to the associated stations. Although the current rating of each cable, obviously, must be adequate for the load, the overriding factor is more likely to be the prospective fault level at the switchboard. As faults may arise at any point in a cable, the cable impedance must not be considered as an alleviation of the prospective fault level except at the far end.

Ring main system

Each end of a ring main system cable terminates in its own circuit-breaker and at each position where a supply is required around the ring, switchgear is essential for tapping-off to either a radial feeder or a transformer. It is common practice to utilize a compact ring main unit. The unit to facilitate this consists of a central circuit-breaker or switch-fuse with wing oil switches. Although it is usual to joint the ring cables to the oil switches with the supply off the central unit, some advantage may be gained by connecting the cable to one of the oil switches and out through the oil circuit-breaker (OCB); this gives protection on each section of cable throughout the ring, but since the protective device has to be capable of carrying the full load current of the ring, the protection afforded to each individual transformer connected to each ring main unit is reduced. With the ring cable taken through both oil switches, each feeder teed-off is protected, but protection for the complete ring is only available at the main switchboard.

The alternative method requires suitable grading of the ring main unit protection as it forms part of the ring and, also, transformer or feeder protection are not as closely related to their requirements. Figures 4.1 and 4.2 depict the methods described. On an extensive system the ring cable may be taken through one or more further high voltage switchboards from which supply is given to additional radials, rings or transformers, in which case circuit-breakers are incorporated in the board for the ring cables. Again, these circuit-breakers must have discriminating and possibly directional protection.

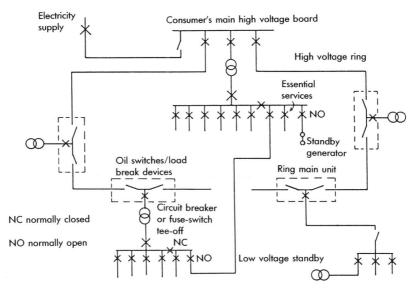

Fig. 4.1 Standard method of ring main connection provides local transformer protection.

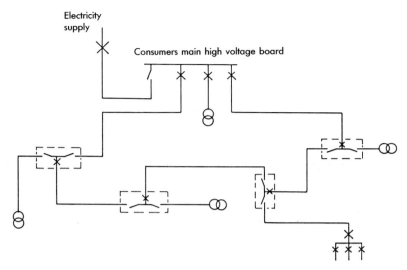

Fig. 4.2 Alternative method of ring protection provides ring cable protection.

Metering installed on the switchgear at the source of supply can only cover the total load on the ring and, therefore, it is preferable to include metering facilities on any substations or intermediate switchboards to provide details of the consumption at each separate location.

Ring main units provide the advantage that, even in the event of a cable fault on the ring, it is still possible to maintain or restore supply to all points by opening the oil switches on either side of the fault. Additional outlets to the ring may also be easily installed without shut-downs, and greater maintenance facilities are available both on the ring feeder circuit-breakers and the ring main unit. In addition, if the main supply switchboard is sectionalized and fed from more than one transformer or alternator, a ring main may be supplied from two different sections, thus improving security.

Apart from fault conditions, it is also possible to operate a ring main in open mode as the preferred arrangement but, for whatever purpose, it must be ensured that neither of the two sections is overloaded.

Load centres

Factory-built assemblies

It is frequently possible to divide commercial and industrial projects into areas of relatively concentrated loads, such as floor-by-floor in a multistorey office block or production areas in a factory, and it should be the objective to locate transformers, distribution switchgear, etc. in close proximity to these areas to keep low voltage cable runs to a minimum. Compact boards, or load-centres, incorporating a transformer are available from manufacturers which incorporate all the facilities that may be required by the consumer and, once installed, need only the cabling to equipment to be completed. Facilities for metering and/or remote control may be included and the disadvantage of separate items of switchgear and distribution boards being located in remote and possibly inaccessible areas is avoided (Fig. 4.3). Low-voltage switchgear and control gear assemblies are covered by BS 5486.

Each load-centre requires a main disconnector, which may be a simple load-break switch, fused switch or circuit-breaker. With the first, protection is essential at the supply end of the feeder while, with either of the alternatives, the rating is chosen to give discrimination with preceding protective equipment. As an example, an accepted rule-of-thumb for high rupturing capacity (HRC) fuses, in the absence of manufacturers' data, is a ratio of 2:1 between fuse ratings (further discussed in Chapter 9) to obtain satisfactory results. Consequently, a 100 A main fuse at the load-centre would be backed up by a 200 A in the supply board and the cable between the two sets of fuses must also be of 200 A rating after taking into consideration voltage drop and applicable correction factors.

Site-assembled designs

In most cases the lowest capital cost is incurred by purchasing separate items of equipment and assembling on site. However, the labour costs involved may offset any savings on capital.

On small site-assembled low voltage load-centres it is usual to install single-core cables between the incoming and distribution gear, preferably in trunking. As there is a limit to the number of cables that a given terminal can accept, it is better engineering to install a length of commercially available three-phase busbar system between the two, each item of equipment then being mounted on the busbar casing and connected directly to the bars. Each piece of switchgear must be suitable for its application, with provision for possible future additions and, preferably, obtained from one manufacturer.

There is an extremely wide choice of designs available for site-assembled load-centres – in price, quality and manufacturer – but, except for very small units, this method is not recommended as, inevitably, it leads to untidy arrangements, alterations and additions with, possibly, dangerous conditions.

The true load-centre consists of a purpose-made cubicle-type assembly

Fig. 4.3 Typical load centre switchboard.

built to the customer's specification; all the required equipment is installed inside the cubicle, with provision for operation on the front panel. Such cubicles are manufactured to standard metric sizes and often consist of a number of elements assembled as one unit. Switchgear manufacturers normally only use their own equipment for load-centres but there are many firms which will buy in equipment to the customer's choice and have the necessary expertise to assemble and test the finished product. With such units, the user/customer has only to provide a suitable base and joint-in the external cabling.

Transformers

Chapter 1 referred to oil-filled transformers, these being the type favoured by the regional electricity companies, but other types are available although, usually, at a higher cost. However, circumstances may be such that cost is not the overriding factor. If transformers are to be installed in occupied buildings (particularly in basements, at roof level, or on the upper floors of a high-rise building) or in dangerous surroundings, dry-type and encapsulated designs or those filled with less flammable liquids than oil (e.g. silicone or Formel) provide many advantages; for example, they have lower maintenance requirements and avoid the possibility of oil spillage, with the attendant risk of fire.

Dry-type transformers, as the description implies, do not have a liquid for insulating and cooling purposes, relying upon air cooling with enhanced insulation for windings (Figs 4.4 and 4.5). The higher operating temperatures usually require greater ventilation facilities than a comparable oil-filled unit. Encapsulated dry-type designs have their windings completely enclosed in resin to provide protection from atmospheric contamination and to minimize interwinding faults. In their infancy problems arose with the differential expansion rates of materials employed leading to cracking of the resin, but this has been overcome (Figs 4.6 and 4.7). Although they minimize fire risks, they are costly, quite bulky and heavy.

The alternative liquid fillings referred to have excellent qualities for insulating and cooling and are very much safer; it is quite possible to retrofill existing oil-filled transformers with the less flammable liquids provided the work is carried out by experts in this field.

An early replacement for oil was polychlorinated biphenyl (PCB) which reduced fire risk but, unfortunately, proved to be highly toxic. The use of PCB was discontinued by manufacturers but some transformers of this

Fig. 4.4 Brush Transformers Ltd dry-type air-cooled-transformer in ventilated enclosure.

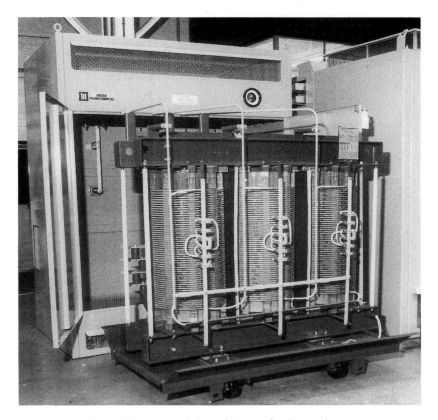

Fig. 4.5 As Fig. 4.4, withdrawn for inspection.

type may still be in use. The reader is referred to BS 171 which covers power transformers.

Packaged substations

Manufacturers developed the packaged substations initially for use by the old supply companies mainly in rural and urban areas, and many of these are still giving satisfactory service. The early type consisted of a weather-proof sheet steel structure, some $10\,\text{m}^2$ in size, with doors on all four sides. They were divided internally into three compartments; the transformer was erected in the middle section with the high voltage switchgear in one end, mounted on the transformer but segregated from it, and the low voltage distribution equipment − usually the interior of a feeder pillar (Fig. 4.8) − at the other, again physically separated. However, there are many variations of packaged substation, e.g. with a centre

Fig. 4.6 Brush 2.5 MVA cast resin (encapsulated) transformer ready for fitting with wheels or into a ventilated enclosure.

walkway, the low voltage switchgear separate, etc. which allow the use of cheaper indoor equipment in an 'outdoor' location. All the equipment was easily accessible through the appropriate doors for operational purposes or for removal. Again, installation was relatively simple, involving only the provision of a suitable concrete plinth and the installation and jointing of cables.

Although the above type was produced specifically for outdoor use, it could equally well be used indoors in factories, providing as it did a completely self-contained substation with or without the housing. The latter, perhaps, lend themselves better to general use as they can be erected in existing suitable accommodation.

Due to the many different requirements of commercial and industrial installations, it may be preferable to purchase a package unit consisting only of the transformer and associated main high voltage and low voltage switchgear. This permits early ordering of the usually long delivery time

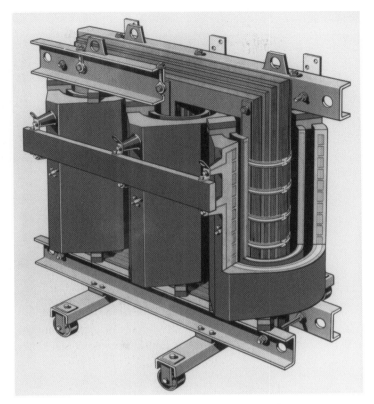

Fig. 4.7 Brush cast resin transformer showing details of construction.

for the equipment (and avoids possible delays) while giving more time to the design stage for the remainder of the installation (Fig. 4.9).

Cable distribution systems

Different cable designs are adopted for both high voltage and low voltage main and submain distribution systems, the choice depending largely on the load, type and requirements of the user. For long above-ground routes subject to large loads, armoured multicore cables are the first choice as they are available with large current ratings, have adequate mechanical strength for suspension purposes, are not easily damaged due to the armouring and have the further advantage that the armouring is usually suitable for use as a protective conductor. The problems associated with hot-compound joints have been eradicated by the development of

Fig. 4.8 Typical low voltage feeder pillar totally insulated complete with removable fuse-operating device providing fault-making load-breaking capability.

Fig. 4.9 Brush 1000 kVA unit (package) substation with high voltage ring main unit and low voltage terminations. This unit is available with or without a weatherproof ventilated housing. Moulded-case circuit-breakers may be provided to replace the low voltage terminations.

cold-jointing techniques and heat-shrink sleeving, both of which are relatively easy to apply and reduce jointing times. Although polyvinyl chloride (PVC) is the most commonly used insulation; it is gradually being replaced by cross-linked polyethylene (XLPE) insulation although there are many alternatives available which are suitable for use in areas subject to arduous conditions such as high temperatures, corrosion and hazardous zones.

Armoured cables are equally suited for direct burial, and installation in prepared ducts or trenches and at high level on tray, ladder rack or cable brackets. With the latter it is wise to consult the cable manufacturer about maximum suspension distances to avoid sagging.

On installations involving many short distances between terminations, single-core cables run on tray provide a more suitable alternative because they have greater flexibility than armoured cables but, in both cases, minimum bending radii must be observed which is obtainable from the manufacturers of the cable. Single-core cables in alternating current systems must not be armoured with magnetic material and, therefore, may require additional mechanical protection; it is also essential to provide protective conductors or, alternatively, to ensure that the supporting metalwork meets the minimum impedance requirement and has sufficient thermal capacity to carry fault currents.

Underfloor ducts in commercial and industrial premises have the advantage that they provide excellent protection for cables and only require sealing at the ends, whereas trenches formed in floors are difficult to seal permanently. However, accessibility is far greater with the latter and, therefore, less problems arise if additional cables are required or a faulty cable has to be repaired or replaced. Although, in Britain, the more onerous requirements related to voltage drop have been removed, the total drop for both multicore and single-core cables must still be such that the operation of plant is not unduly affected, and the correction factors referred to in Chapter 8 must be applied. When the user obtains supply from his own transformer, it is possible to offset the voltage drop effect on the plant by changing transformer tappings, but this could lead to undue fluctuations in voltage as load requirements change and is, therefore, more applicable for relatively steady loads. The majority of transformers include ±2.5 and 5% tappings but, on the smaller types, the tap-change links are usually internal and this necessitates removal of the lid to effect alterations involving shutting down and isolating the unit.

Chapter 5
Main and Standby Distribution

Metering

On the size of generating plant required for the smaller installation where it is improbable that it will be used for long periods, the metering included in the control panel is usually confined to a voltmeter and ammeter with selector switches, an hours-run meter and the necessary instruments to indicate the general state of the prime mover, such as oil pressure gauge, temperature indicator and fuel content.

If private generation is required for the more extensive uses indicated in Chapter 2, i.e. base load or peak lopping, this is obviously inadequate as it provides no information on the user's consumption, whether it is on load at the correct times, or even whether the generator is capable of supplying the imposed loads. Further, if a set is intended for parallel operation with the public supply or other sets or to feed into the mains network, more extensive metering facilities are essential as, usually, the regional electricity company's metering only measures supply into premises.

Under these circumstances the end-user should install, as a minimum, summation metering on the generator main switchgear and, more essential, instrumentation and switching facilities to enable synchronizing to be effected. Synchronizing requires, at least, that polarity and phase rotation have been checked at installation, and that frequency and voltmeters are available on both the live busbars and the generator supply side as these are necessary to be able to exercise close control when synchronizing; the two sets of instruments must be in close proximity to each other. A more efficient method of obtaining parity between two sources of supply is by use of a synchroscope which may be either a miniature induction motor or a dynamometer type. This instrument has only the one indicating needle which revolves in one or other direction as a machine is brought up to speed. At synchronism the needle remains at a marked position on the dial and the generator circuit breaker is then closed manually or automatically via a relay included in the synchroscope circuit. This latter is preferable but more costly, however, it removes all possibility of human error and is certainly a facility that should be incorporated with large generating sets.

As there are greater limitations on private generator loadings than on public supply, it is also an advantage to install maximum demand indicators or peak load annunciators and, in the interests of economy, power factor metering.

One problem that can arise with metering on paralleled supplies is that standard summation meters are bidirectional and, therefore, will reverse direction according to whether the supply is from the electricity supplies into the installation or from the generator into the mains. Consequently, special precautions must be taken to ensure that meters cannot reverse, either by the inclusion of stops on the public supply meters or reverse power bypass relays to take them out of service when supply is in the 'wrong' direction. However, this is normally the responsibility of the regional electricity company rather than the consumer, although any additional cost will almost invariably be borne by the latter.

Although the induction type of meter presently in use worldwide has given good service, it is susceptible to both electrical and mechanical failure and, consequently, must be serviced from time to time. The UK supply industry summation metering must, in fact, be withdrawn at intervals for testing and recertification to comply with legal requirements. At the present time, electronic metering is being produced and installed in certain areas and it is anticipated by meter manufacturers that, in the coming years, this will be the standard type in use. The anticipated advantages of electronic metering are that remote reading and inclusion in computer recording at central monitoring consoles will be feasible.

Essential and nonessential loads

Every commercial and industrial load is capable of being divided into two distinct categories, one of which may be switched off during emergency conditions without causing danger to personnel and plant while the other must be maintained for at least sufficient time to allow escape and, if possible, to shut down any plant which might suffer serious damage if left running or, alternatively, endanger those who would be attempting to deal with the situation. The latter, in the IEE Wiring Regulations, are generally referred to as the safety services, that is, the essential loads.

As each installation provides different requirements, it is not possible to be completely precise regarding all aspects, but services such as fire alarms, emergency and standby lighting, data and computer installations and security systems are almost universal while, in buildings equipped with sprinklers or hoses for fire fighting, any associated pumps and controls are classified as essential (Fig. 5.1).

In many countries there are statutory requirements governing essential services such as, in the UK, the Health and Safety at Work Act, in addition to which many local authorities apply further conditions before a building is allowed to be used.

The major consideration is that essential services must not be disrupted, by any cause, until the last possible moment. It is therefore necessary to ensure that the source of supply for them is independent of that for the nonessential services, by the use of either batteries or generators, and that no part of an essential system is installed in such a location that it is likely to be damaged by other sources including nonessential services or known hazardous areas. The British Standards Institution's Codes of Practice for Fire Alarm Systems (BS 5839) and Emergency Lighting (BS 5266) specifically require all wiring to be segregated from the normal nonessential wiring either by suitable barriers or by distance. Although it is possible to utilize a generator for standby purposes in addition to the safety supplies, it is also essential to ensure that, in an emergency, only the safety services remain available; consequently, switchboards obtaining

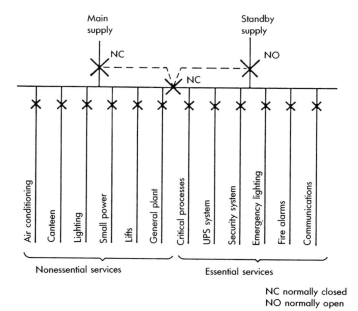

Fig. 5.1 Single-line diagram of consumer's switchboard with essential and nonessential sections.

supply from both main and secondary sources must be equipped with control devices that will disconnect nonessential supplies. This is of particular importance when different supplies operate in parallel.

Although all electrical installations should be subject to periodic inspection and testing, this is of particular importance with regard to essential services. In the case of fire alarm systems and emergency lighting it is preferable to include constant monitoring equipment and, with emergency lighting, to include easily accessible test switches. If emergency luminaires cannot be reached easily, it is possible to include remote testing devices for the circuitry or built-in photocell-operated switches which are initiated by the beam of a torch.

Cable distribution

For practical purposes cables, whether for distribution, subdistribution or final circuits, are installed in groups to avoid a multitude of routes; this enhances appearance in exposed runs, and minimizes the cost of fixings (Fig. 5.2).

Starting at the main switchboard, the distribution cables, which are normally the largest, are usually installed in builders' work ducts or on ladder-racking or heavy tray and, although they are not very flexible, usually present little serious difficulty when so grouped as space is not at a premium. It is advantageous, however, to plan the individual routes carefully to avoid unnecessary crossovers.

As an installation extends, it is usually the case that the number of subdistribution cables increases and may take the form of multi- or single-core cables installed on tray or in trunking (on a.c. circuits single-core cables must not be armoured). When single-core cables are the choice, all cores of each circuit, including the neutral, should be strapped together. The size of tray or trunking installed must be adequate for the number of cables (and possible future circuits), bearing in mind the space factor where relevant, and to allow segregation of safety services as mentioned in the previous section. Advantage of spacing may be taken to separate individual cables and allow the minimum sizes of cables to be used due to the increased cooling facility. Again, with cables on tray, preplanning of routes is necessary to avoid crossovers, while in trunking barriers are available to provide segregation and cable retainers facilitate cable installation.

Final circuits, if grouped in large numbers, should preferably be installed in trunking, branching off in conduit or mini-trunking to individual cir-

Fig. 5.2 BICC power distribution cables with supportive rackwork at the John Player Horizon factory, Nottingham. Note the individual clipping of the cables.

cuits. It is essential in this case to take careful consideration of grouping to avoid derating by high grouping correction factors, and also to ensure that associated circuit cables are strapped together and identifiable throughout their length if in trunking or at each end, at least, when in conduit.

If safety circuits, particularly fire alarm and emergency lighting cables, are included in the same trunking, they must be mechanically segregated from all other services either by metal barriers or by the use of copper-sheathed cables for the essential services. When essential services are extensive, it is preferable to install them independently and, if possible, on different routes.

On extensive installations, cable systems are an efficient and relatively economical means of transmitting large blocks of power between points, but benefit may be lost if separate cables are installed for each essential service, as in many cases these are lightly loaded and it may not be possible to size down the cables to the actual current rating required. Additionally, with variable loads, separate cables do not give the full advantage of diversity. However, these problems can be overcome by supplying essential services from satellite switchboards rather than from the source of supply. For those services normally incorporating battery/rectifier/charger equipment this presents no difficulty and, as in the case of nonmaintained emergency lighting which only operates on failure of supply, gives the advantage that it relates more to the area in which it is located.

By adopting such a method full advantage is obtained from diversity and loadings on the main distribution cables, and, in addition, a number of smaller essential services supply units reduces the possibility of widespread failure inherent in the installation of a central equipment.

Split distribution systems

On many installations circumstances may exist which call for rather more than the segregation of circuits for compliance with the applicable IEE Wiring Regulations, and it then becomes necessary to install split distribution systems (Fig. 5.3).

In the most simple case, a domestic or similar small installation, this is required when a two-part tariff applies − lighting being charged at a different rate (normally higher) than the power − and is effected easily by installing one distribution board and main-switch for each service. The regional electricity company then installs two meters and connects the meter tails into the boards separately. Similar arrangements are made when special tariffs apply to part only of an installation, as in the case of the Economy 7 referred to in Chapter 1.

A further situation which calls for a split distribution system arises from the increased emphasis on the use of RCDs. The IEE Wiring Regulations state that where it is likely that a socket-outlet may be used for equipment outdoors the socket-outlet must be protected by an RCD but, as mentioned elsewhere, inherent earth leakage may be present which could cause nuisance tripping. In these circumstances it is good practice to install separate systems and, for domestic or similar installations, consumer units are available which provide this facility.

In commercial and industrial installations similar arrangements may be required, again where certain tariffs are applicable to a part of the

Fig. 5.3 BICC power distribution cables supported by Vantrunk cable ladder with separate traywork for essential or ancillary services.

installation such as floodlighting or advertising signs. In view of the increasing loads being installed for equipment such as air conditioning, refrigeration, lifts and motive power – types of equipment which may impose sudden loads at varying intervals – it is of advantage to install separate systems from those which are less subject to variations such as lighting and, particularly, sensitive electronic equipment. Each system is supplied from as near as possible to the origin of the installation to minimize the effects of voltage fluctuations.

Again, when designing an extensive installation, there may be advantages in splitting it, at the source of supply, into two parts: essential and nonessential services. The first includes all safety services such as fire alarms, emergency lighting, security, communications, supplies necessary for specific processes to avoid damage to plant or materials, and the UPS system described in Chapter 2.

This is achieved by installing two busbar sections in the main switch-board, normally connected together through a bus-section switch or circuit-breaker. The main incoming supply is connected to the bars on the nonessential section and a standby generator, or alternative public supply if the authority permits such, to the essential services section. A split installation of this type is more effective if the bus-section device is designed to open automatically on loss of supply and also start the generator or close the alternative supply circuit breaker. However, the designer must take into consideration the quality of maintenance available to the end user, as although automatic change-over simplifies operational procedures it also increases the maintenance requirement if electromechanical relay equipment is used to effect the necessary action. Static switching is equally as effective and, in the event of failure, may reduce rectifying procedures to no more than the replacement of a printed circuit board (PCB) which, although not eliminating the problems of a shutdown, will reduce the down-time.

Chapter 6
Busbar Distribution Systems

For installations requiring a minimum amount of distribution, such as housing, small shops and offices, a single-phase service is usually adequate and the wiring is carried out in twin and circuit protective conductor (CPC) PVC insulated cables (BS 6004) or single-core cables in conduit or trunking run in lofts and floor voids (when available). With heavier loads for large commercial premises and factories, there are more onerous requirements and, consequently, distribution systems are far more substantial; for main and submain feeders, armoured cables are well recognized but, where extremely heavy loads are to be considered, busbar systems have much to recommend them. Additionally, the latter are particularly suitable when tap-offs are needed, for example at each floor of a multistorey block or for machine shops, and many different types are available.

Horizontal busbars

The two main types of prefabricated horizontal busbar are those containing either bare copper bars supported at suitable intervals by insulated carriers or totally insulated bars, and both are available with a number of variations. Each type is manufactured in ranges of standard lengths allowing almost any total run to be obtained although, in an extreme case, the majority of manufacturers have the facilities to provide a one-off length. Whichever type of busbar is selected, there are no problems over mechanical protection as this is provided in the overall metal casing while, in addition, the strength allows the minimum of support which, properly planned, gives the designer/installer the opportunity to make full use of structural steelwork or, if supported from a wall or concrete ceiling, requires far less drilling for bracketing. It is frequently possible, also, to carry out much of the assembly work at floor level, thus reducing the amount of scaffolding or ladder-work usually associated with high level installation work (Fig. 6.1).

If long lighting runs are required, busbar trunking is available complete with outlets spaced according to the requirements of the purchaser, and

Fig. 6.1 Power Centre Holdings Ltd. type RPN plug-in busbar trunking installed at Land Rover, utilizing plug-in tap-off fuse boxes mounted on the underside of the trunking.

with built-in provision for luminaire suspension. One such type is equipped with socket-outlets which, again, reduces the amount of work at high level as luminaires are simply completed on the ground with a flexible cable and plug and then hoisted on to the suspensions. For the lighter loads the bars usually consist of PVC insulated single-core cables supported within the trunking, rather than copper or aluminium bars.

For heavier loads which require numerous tap-offs, busbar assemblies are available complete with outlets consisting of terminal blocks, allowing simple connection of cables to feed machines, or with plug-in units, the latter being complete with isolators or fusegear as required. Some form of protection at the tap-off point is strongly advised, otherwise the rating of the machine cable feed must match the busbar rating since a fault at the load could cause a complete burn-out of the cable.

One further use of a busbar assembly is between a transformer and the main low voltage switchboard to avoid having to install two or more cables for each phase. This is not usually done, however, for the size of installation under consideration as it would entail the use of nonstandard cable terminating boxes on both transformer and switchgear, modifications which, although not presenting difficulty for a manufacturer, would probably entail additional costs which might outweigh any benefits.

Vertical risers

The construction of vertical risers is similar to that of horizontal busbars except that access to the busbars is usually possible along the whole length of the riser by the inclusion of relatively easily removed covers. Such risers are particularly suitable for use in high-rise buildings where it is required to tap-off at each floor and accessories are provided to facilitate this, complete with protective devices such as fusegear or moulded case circuit-breaker (MCCB) equipment (Fig. 6.2).

Fixing brackets are incorporated in the structure of the metal casing and, as with horizontal busbars, the number of fixings is minimized. Although it is common practice in prestigious buildings to install vertical risers in ducts built into the structure, usually provided for other services also, it is preferable for them to occupy separate ducts as, normally being out of sight, it is quite possible for a water service, for example, to develop a leak which may not be immediately noticed and may cause damage to the electrical installation. Many risers, in any case, look quite presentable and, if installed through a staircase or similar location, are unlikely to detract from the general appearance of a building. If an architect or the owner of a building is persuaded to accept an open position, any switchgear or metering connected to the riser should be enclosed in a cupboard large enough to accept the equipment and provide adequate room for maintenance or for repairs to be effected.

The most simple form of riser consists of bare copper or aluminium busbars mounted on insulators fixed at suitable intervals to the building structure. To reduce the number of fixings it is possible to obtain horizontal sets of insulators mounted on steel channel or angle iron brackets which are then rag-bolted to the structure where required. This method simplifies the operation of ensuring vertical alignment of the busbars as each bracket assembly is jig-drilled at the manufacturers, and it is then only necessary to align each assembly rather than each insulator.

As this type of riser is not of the enclosed type it is more suitable for industrial use, where appearance is not as important as in commercial premises, although it is installed in some of the older, large office blocks built when space was not as costly as at the present time.

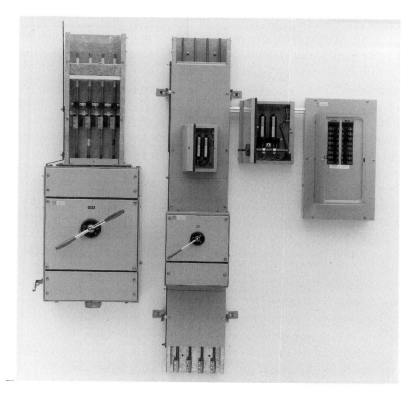

Fig. 6.2 Power Centre Holdings Ltd. type RRM rising main busbar trunking
showing DF7 fuse-switch incomer, fuse tap-off box cover-mounted and Century C
switch-fuse side-mounted feeding a Powermark minature circuit-breaker (MCB)
distribution board.

The disadvantage associated with bare risers is that the spacing between
busbars is normally greater than with enclosed equipment as there is the
greater danger of faults arising in dusty or humid atmospheres or, in some
cases, from birds settling on the brackets etc. Obviously, being unenclosed, it is essential to provide some form of structural barrier around
the equipment to avoid accidental or unauthorized contact but, depending
upon the location, this may be no more than close-mesh screening having
an IP2X category (index of protection, BS EN 60529: 1992).

Unless the riser is installed within a fire compartment, it is necessary to
provide barriers at floor levels etc. which are both insulating and fireproof.

Busbar material is available in either rectangular- or round-section,
although rectangular is more suitable as it enables tee-offs to be effected
easily by means of clamps or drilling and bolting and, should it become

necessary, to increase the capacity of the riser by adding further sections to each bar. The rod section requires shaped connectors for both increasing the capacity and providing tees, these being similar to the line-taps used for connections to overhead lines.

The assembly of this type of riser is a site operation and, as the busbars tend to expand and contract under the effects of load and changes in ambient temperature, consideration must be given during assembly to the inclusion of flexible sections on long risers and at the supply end. If supply is provided via multicore cables, flexibility is obtained by terminating the cable in an end-box and extending to the bars with single-core cables. The necessary degree of flexibility in manufactured risers is provided during assembly by purpose-made flexible straps.

For enclosed busbar risers to comply with the 16th Edition of the IEE Wiring Regulations, fire seals must be installed inside the enclosure as well as round the outside where they pass through a free barrier.

Mini busbar systems

The demand for low wattage multi-outlets in many fields has given rise to the development of various forms of mini busbar system which remove the necessity for the unsafe practice of installing adaptors into lampholders (which are no longer recognized by the IEE Wiring Regulations) and numerous trailing flexes. The use of mini busbars provides an inconspicuous and convenient method of supply for groups of luminaires or light load equipment such as typewriters, calculators, VDUs, etc. with the additional benefit that alterations to installations are easily carried out without major disruption. Such facilities (usually referred to as lighting trunking) are particularly suitable for display and feature lighting in housing, offices, shop windows, exhibition halls and art galleries, to name but a few examples (Figs 6.3 and 6.4).

The busbars are totally enclosed on three sides with return edges on the fourth providing a full-length slot normally covered by a flexible shield; special connectors are provided, for terminating leads from equipment, which are inserted through the cover and securely locked in position by a cam-type arrangement on the connector which also forces contacts on to the busbars. For display lighting, luminaires are available which incorporate a connector in the construction and, consequently, can be plugged directly into the lighting trunking.

Although this type of busbar is extremely safe against inadvertent access to the live bars, there is the possibility that thin metal objects or

Fig. 6.3 Office lit with conventional track lighting. (Courtesy of Hugh King, Thorn EMI Lighting.)

small prying fingers could defeat the safety features and, therefore, as with all electrical equipment and installations, due care should be taken with regard to location or, alternatively, it should be used for extra-low voltage circuits. Additionally, it is preferable not to install it in damp or dusty atmospheres.

Other types, with higher current ratings and used in conjunction with the standard 13 A outlet and plug, are available which are more suitable for general use in offices and commercial installations. One such is the MK Powerlink system which consists of uPVC trunking containing three 63 A busbars, single-phase and earth, which is designed for surface mounting at skirting or dado level. The basic section is divided into separate compartments to provide facilities for telecommunication or data wiring completely segregated from the busbars, with the additional advantage that a matching extension section is available which interlocks with the basic assembly (Fig. 6.5).

Fig. 6.4 Rackhams, Birmingham, using Thorn EMI track and low-voltage spotlight. (Courtesy of Hugh King, Thorn EMI Lighting.)

Should the user so wish, Powerlink can be installed simply in trunking form, as the busbar assembly is a self-contained unit which is easily installed at a later date. Similarly, outlets may be added or removed as required for although, as mentioned previously, they are of the 13 A type, the back of each outlet is of a modified pattern which enables it to be plugged into the bars after removal of the lid.

As with other systems, accessories are available to suit every requirement for an installation and, additionally a 32 A two-pole switch which may be used to provide supply to the busbars or to external equipment, and again simply plugs into the bars. The range of accessories also

Fig. 6.5 MK Powerlink, a distribution system which operates on a simple busbar principle. It also offers the facility for running ancillary services such as telephone and computer cables in the same trunking system but completely segregated.

includes adaptors which allow mini-trunking to be connected to the surface track for small extensions.

A similar form, but with 30 A rated busbars, is produced by Electro-patent Ltd under the trade name of Multipoint. As the bars are formed to accept the pins of 13 A plugs at 300 mm intervals, the method of construction is relatively simple as the outlets are, in effect, simply faceplates premounted on the trunking. Multipoint is produced in five standard lengths varying between 0.3 m and 3.6 m complete with secondary ducting for the segregation of other services in a parallel-sided section only but, as the outlets are mounted at 90° to normal, may be installed at either dado or skirting level without creating problems for flexible cables attached to plugs.

Track systems

During the 1970s, there were developments, of an unconventional nature, with track systems (described below) which eliminated some of the problems associated with normal fixed-wiring installations. These systems are designed to avoid excessive relocation of concealed fixed-wiring or to provide multiple and closely associated socket-outlets.

Undercarpet systems

Particularly in areas such as large open-plan offices, extensive floor-duct systems may not always be possible (see Chapter 11), and the undercarpet system, which originated in the USA is increasing in popularity. Typically, this consists of wide but thin conductors, laid side by side, sandwiched between layers of strip insulating material; over this is laid a full-width copper strip protective conductor, in turn covered by a similar stainless steel sheet. Accessories are available which allow jointing to normal fixed wiring, for floor-mounted outlets, for jointing at cross-overs and almost any conceivable requirement. In addition to power circuit wiring, the system is available with additional conductors for communication and data transmission facilities but, unless the total requirements are known in advance, a duct system is more versatile and has far greater capacity. As the complete assembly is extremely thin, it is very well suited for installation beneath floor coverings although, to obtain the maximum benefit, it is preferable for these to be in the form of tiles so that any rearrangement of furniture will only require the minimum removal of floor covering for the repositioning of cable and outlets.

This system (Versa-Trak) is available from Thomas and Betts Ltd.

The original American system appeared to have been specifically designed for a 100 V 60 Hz supply, whereas that now produced in the UK has been improved for use on a higher voltage.

Multipoint outlet system

As mentioned earlier, the increased demand for outlets in domestic, commercial and laboratory installations creates difficulty when wiring is concealed beneath plaster, flooring, etc. whether it is twin and CPC cable or PVC insulated cable in conduit or trunking. To meet this demand, a busbar system has been marketed which allows special plugs to be inserted directly into the bars at frequent intervals; the spacing of the outlet points, if different from the standard, must be arranged with the manufacturer in advance to enable the outlet aperture shielding to be provided at the appropriate positions. The protective shield opens upon insertion of the plug which, when fully home, is turned through 90° to bring the blades on the plug into contact with the busbars and lock it in position. The plugs are completely different in shape from the standard 13 A type but they have the same facility for self cleaning and fuse protection.

The system, Eletrak, is also available with provision, by means of flexible connecting links, for several lengths of trunking to be joined together to provide an extensive system, although protective measures must be carefully studied to ensure that overload and fault conditions are adequately catered for.

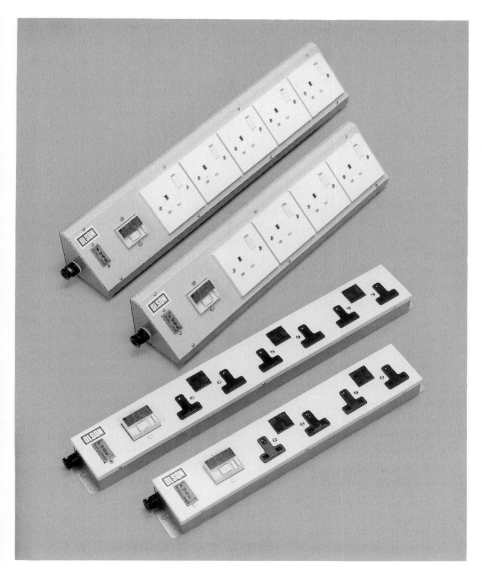

Fig. 6.6 Multipoint socket-outlet units with RCD and fuse protection, suitable for wall or bench mounting. (Courtesy of Olson Electronics Ltd, London.)

In addition to the above, several systems are obtainable which utilize BS 1363, 13 A plugs and, therefore, avoid the inconvenience associated with different types of plug-in installations already equipped with standard ring circuits. These are usually provided in relatively short lengths complete with up to six outlets and, consequently, are quite suitable for areas such as worktops in kitchens, test benches, laboratories, schools, computer desks and workshops where several small tools or test instruments are regularly used in close proximity to each other.

Although the number of outlets is restricted, they are available complete with coiled cable and 13 A fused plug and, therefore, are easily trans-ferred, if required, to a different location. An additional safety feature provided by some manufacturers is a built-in 30 or 10 mA residual current circuit-breaker (RCCB).

Different profiles are produced such as rectangular (approximately 100 mm by 50 mm) and console shape (100 mm by 100 mm) which make them equally suitable for fixing at skirting or higher level or against the angle of a working surface and wall. On the console type, the outlets are situated on the angled face which avoids sharp bends on the equipment flexibles at the plug (Fig. 6.6).

Chapter 7
Cable Types and Applications

Regional electricity companies' practice

Since the early 1990s, the continued growth of the supply industry world-wide has provided the greatest incentive for manufacturers to produce better and better types of general purpose cable, while such industries as chemical, oil and petrol, mining, shipping and textiles have created increased demands for special cables for use in hazardous areas and adverse temperature conditions. This has resulted in different materials being used for conductors (partly for reasons of economy), insulation and armouring from those that were originally used, even though many of the early types of cable stood the test of time and, indeed, are still in operation. The modern cables, in addition to possessing many advantages, also, unfortunately, demand completely different techniques in both handling and installation, particularly with regard to terminations. These developments are further discussed in the following sections.

High voltage supplies

Until approximately the middle of this century, the most common type of high voltage cable (up to 11 kV) in use consisted of stranded copper conductors, oil-impregnated paper insulated, steel wire or tape armoured over a lead sheath, and served overall with bitumen-impregnated jute or similar material. The main variations were related to the form of the conductor which were available in circular or segmented cross-section and, from some manufacturers, in solid rather than stranded copper. Some disadvantages of this type of cable are that: firstly, jointing procedures are very complicated and time consuming, particularly in adverse weather conditions; secondly, after a long period of use the serving tends to rot, lead sheathing develops cracks and insulation impregnation dries out, resulting in moisture penetrating to the cable and increasing the possibility of faults developing and; thirdly, oil impregnation tends to migrate.

Present cable practice involves the use of aluminium conductors made to international metric sizes (this also applies to copper conductors), the replacement of lead sheathing by aluminium and PVC oversheath. Impregnated paper is still used extensively for insulation on 11 kV cables as the possibility of drying-out has been largely overcome by the use of nonhygroscopic PVC and similar materials for oversheaths, these being less susceptible to deterioration (Fig. 7.1).

Although, initially, the use of aluminium sheathing presented difficulties due to the inflexibility of the material, various designs were introduced, such as corrugated formation, which provide adequate flexibility for the purpose.

The jointing and terminating of cables had been greatly simplified by the development of mechanically compressed ferrule connectors and lugs (Figs 7.2 and 7.3) and the use of cold compounds and chemically reactive materials for filling which, in effect, produce encapsulated joints which are almost indestructible and provide barriers against moisture creepage.

Low voltage supplies

Similar developments have taken place for low voltage cables and low voltage/regional electricity companies in the UK almost invariably install low voltage cables in which neutral and earth are combined in one conductor acting as an oversheath for the phase conductors. The phase conductors are insulated with cross-linked polyethylene (XLPE) around which the combined conductor is embedded in unvulcanized rubber, the whole then being encased in an extruded PVC sheath. This type of cable, produced under the descriptions of 'Districable', 'Consac' or, the most commonly used, 'Waveform', has the advantages that it is relatively low in cost, is sufficiently flexible, incapable of moisture absorption and easily jointed with compression ferrules, heat-shrink sleeving and encapsulation (Fig. 7.4).

Fig. 7.1 BICC three-core 11 kV belted PIAS cable with corrugated aluminium sheath as used by UK supply industry.

Fig. 7.2 Typical cable jointing ferrules and terminating lugs for application with compression tool.

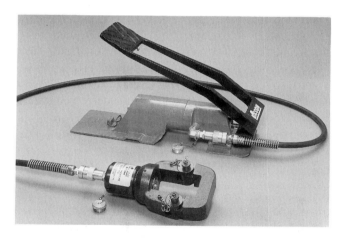

Fig. 7.3 BICC compression tool type BICCON.

Generally, the above types of cable and the construction materials are covered by British Standards (most of which have been harmonized with International and European Standards), and, additionally, standardization requirements have been intensified by the introduction of the British Approvals Service for Electric Cables (BASEC) certification scheme which

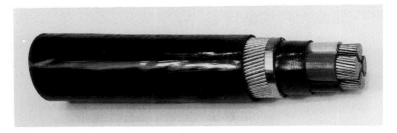

Fig. 7.4 BICC four-core XLPE insulated, extruded PVC-bedded, galvanized steel wire-armoured, PVC-oversheathed 600/1000 V cable to BS 5457: 1977.

guarantees that any cable so certified is to the highest quality. It may be of interest to the reader, however, to know that many British cables have been produced over the years which, although classified as 'standard', have far exceeded minimum requirements. This is particularly the case with so-called tropicalized cables for use in high temperature situations in industry or normally hot countries.

Consumer installations

In the natural course of events, cables produced for the supply industry are also used by the private consumer for general application but, as mentioned above, situations arise in industry in particular which call for cables suitable for high temperature or for corrosive-proof and flameproof installations, conditions which are exceptional for the supply industry. These special types are obtainable from all major producers.

High voltage installations

Very often in the supply industry, the prospective fault currents, which are of the order of 13.1 kA and higher on 11 kV systems, govern the cable size, rather than the much lower load currents. On the other hand, consumers are usually located some distance from major supply sources and, therefore, due to the added impedance of the systems, have less onerous conditions to cope with. This usually permits the use of lower rated cables, i.e. more compatible with loadings, although this does not remove the necessity for a consumer to obtain the fault level from the regional electricity company.

Reference has previously been made to the simplification of jointing methods for modern cables (Fig. 7.5) but, despite this, joints and terminations do introduce potentially weak points in a cable run and, conse-

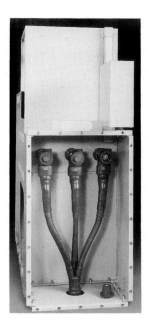

Fig. 7.5 Yorkshire Switchgear 11 kV termination for plastics-insulated cables using compression lugs and moulded insulating shrouds.

quently, only the highest quality materials should be employed and manufacturers' recommendations followed when jointing or terminating. For the specialized cables this is even more important as, for example, those installed in flame proof (FLP) zones require approved (certified) joint boxes which are of much stronger construction than those for general installations to prevent, in the event of a joint failing, sparks or flames emanating from the joint box into the hazardous area.

In such cases it is preferable to avoid paper-insulated cables and install those employing cross-linked polythylene (XLPE) or ethylene propylene rubber (EPR) for the higher voltages while, for low voltage distribution, in addition to XLPE and EPR, specially enhanced fire performance PVC-insulated cables are available. BICC produce a large variety of special-situation cables in the 'Flambic' range but comparable types are available from all major cable manufacturers.

Low voltage distribution

For commercial and industrial installations the most commonly used low voltage cable is the PVC-insulated, armoured, PVC-sheathed type

(PVCSWA and PVC sheathed) with copper conductors, although aluminium conductors are available and are popular as a result of the price differential between the two materials. Initially, aluminium did not find favour with installers due to the fact that many switchgear terminations were of the clamp type and the material tended to creep, from the effects of pressure and heat, causing the terminals to become loose; solid aluminium conductors appeared to suffer more than stranded ones in this respect. A further disadvantage of aluminium is that it has a lower conductance than copper so that, for a given rating, a multicore cable with aluminium conductors and armour is much larger than the copper equivalent, a 300 mm² aluminium cable, for example, being approximately equivalent to a 185 mm² copper one under the same conditions of installation. However, one big advantage with the strip aluminium armour is the good earth return. When aluminium was first introduced, switchgear terminals were mainly related to copper sizes, this presented serious problems which have since been overcome by the use of necked compression ferrules. These also eliminate any requirement for soldering, which also proved troublesome due to the electrolytic action between dissimilar metals.

As in the case of high voltage cables, different materials are available for the insulation and sheathing of low voltage distribution cables which enable them to be installed in abnormal conditions and, again, it is preferable to discuss the requirements with manufacturers.

Reference has been made to horizontal busbar systems (Chapter 6) which are suitable for heavy current situations and simplify tee-ing off to intermediate loads. The advantages of a cable installation are that, on a long route on tray, walls, etc. a continuous run of cable is possible which, although requiring more support than busbars throughout the length, avoids the large and costly labour-intensive amount of assembly work necessary which, in turn, eliminates the possibility of joints breaking down. Further, although many accessories are available to enable busbars to deviate from the straight run, these add to the number of joints and the assembly entailed while cables can easily be bent to shape, particularly when only small changes in route are required by unforeseen obstructions. Finally, as so often happens during the construction of a building, major changes in route may arise which, in the case of busbars, can be difficult to accomplish or can lead to delay while further accessories are obtained. Such changes create far less difficulty with a cable installation provided that the route is not extended beyond the length of cable in hand. If the changed situation demands the purchase of further cable, however, it is possible that the additional cost will be out of proportion to the required length, as manufacturers tend to charge over the basic price for cut lengths. Despite this, the overall cost of a cable run is likely to be considerably less than a busbar installation. Any increase in length of the original busbar system or a cable run would, of course, necessitate the design being checked to ensure that the safety of circuits was still satisfactory.

Circuit wiring

Enclosed cabling

Except when armoured cable requires further mechanical protection, such as at low level or underground, it is not necessary to install it in an enclosure, which may, in any case, add to the difficulties unless split ducts are used. When ducting is essential, it is preferable to lay it in an open trench where possible or, if boring techniques are used, provide frequent pull-in pits along the route to facilitate the later installation of the cable. These installations are normally for the heavier loads so each cable should be run in its own duct and drawn in with appropriate equipment at a steady rate. For the heavier cables, this equipment usually consists of a capstan, wire hawser or nylon rope, and a suitably sized sock which is self-tightening over the cable end.

If an installation requires a number of armoured cables on the same route across, say, a factory floor, an open duct is formed during the groundworks in which racks are installed for the cables. After the work has been completed, concrete lids or checkerplating are placed over the full length of the duct.

For the smaller cables, enclosures such as conduit and trunking, referred to in Chapter 11, are employed and PVC-insulated, with or without sheath, single-core cables (to BS 6004) installed following completion of the conduit/trunking system. As these cables are usually installed in relatively large groups, care must be taken to avoid overheating and to provide identification of the different circuits. The correction factor for grouping is discussed in Chapter 8.

Tray and ladder rack

As tray provides continuous support, unless mounted on edge or in vertical runs (when adequate strapping or clipping is essential), the mechanical strength of supported cable is not as important as with ladder-racking or structural support methods. Consequently, tray is eminently suitable for the smaller unarmoured cabling while racks and structural support, except for short lengths, call for armoured cables as they provide the necessary strength to avoid sagging between supports. Both tray and ladder racks can be provided with accessories to facilitate changes of route, and as PVC and similar insulating materials are nonmigratory (unlike the older types of impregnated cables) they provide no difficulty in this respect on vertical runs.

An alternative to the above cables is mineral-insulated copper-covered cable (MICC) which has the advantage of a higher temperature rating than PVC if not exposed to touch (105°C against 70°C). It has the disadvantage, however, that the insulation, magnesium oxide, is absorbent and it is essential that cut ends are not left unsealed. It is equally important that joints and terminations are correctly formed in accordance with the manufacturers' instructions, although recent developments are claimed to overcome this problem (Fig. 7.6). Reference should be made to BS 6207 and 6081.

Domestic systems

The smaller domestic type of installation is adequately catered for with twin and CPC PVC cables or single-core PVC cable in some form of enclosure; the installation of the first is less labour-intensive than conduit work although the second provides better mechanical protection. Due to the amount of space that is occasionally available between floorboards and ceilings (modern construction methods include solid floors) and in lofts, installation is relatively simple and protection is rarely necessary for horizontal runs. Where droppers are required for switches and wall-fittings, however, it is essential to provide oval conduit or capping over the cables. Although PVC has a much longer anticipated life than the previously used rubber covered cables, MICC is a suitable alternative which has an even longer life. It is not commonly used on very small installations except, possibly, for exterior lighting or feeds to remote buildings.

At the present time a considerable number of domestic refurbishment programmes are in progress for which, to avoid undue disruption and damage to existing floorboards, plastering, etc. a number of enclosed surface systems are available (such as Gilflex Homeline HL2512) which incorporate mini-trunking, dado- and cornice-trunking. For each system, accessories are available for accommodating different types of outlet and for negotiating corners, doorways, etc. A correctly designed installation is

Fig. 7.6 BICC Pyrotenax mineral-insulated cables installed on Vantrunk cable ladder at Glaxo (UK) Ltd, Annan.

effective and relatively inconspicuous, although even where obvious, such as across ceilings, it presents an aesthetically pleasing appearance.

Communication systems

The increase during the latter half of this century of communication systems for TV, radio, computers, data processing, automated plant control and similar, has produced a demand for different types of cable from those used for power systems as, among other requirements, they have not only to be durable but also to be protected from external interference and capable of transmitting a multitude of signals. For these purposes, the following types of cable are now produced.

TV and radio

The main requirement in a consumer's installation, with the exception of the transmitting authorities, is for aerial cables to receivers or between remote screens and cameras and a central control station, all of which must be capable of handling radio frequency signalling and be interference free. As the power and currents involved are extremely low and frequencies high (up to 100 MHz in some cases), insulation and current-carrying capacity are less important than impedance and capacitance.

This type of cable is produced in coaxial form, the central solid or stranded conductor being either plain, tinned, silver-plated copper or, alternatively, copper-covered or silver-plated copper-covered steel. To this is applied an extruded or beaded polyethylene insulant followed by a braided screen of similar material to the main conductor. The whole is then sheathed with extruded PVC or nylon.

Computer data transmission and control systems

Cables required for data transmission and control systems fall broadly into two groups: those that are required between the computer and the outstations and those used between the machine and the associated peripheral equipment.

Generally, the field wiring is of similar construction to that described except that it is multicore and may have screening applied to each core, to each pair or, simply, overall. The range of cables is so great that it would be impossible to describe them all in the space available and, as the requirements of each computer manufacturer differ with regard to the required characteristics, anyone contemplating this type of installation should be advised by them or seek the assistance of the major cable firms (Fig. 7.7).

One of the most popular cables in use for peripheral equipment is the ribbon form which is also produced as multicore cable, with and without screening and various types of insulation. Although ribbon cables are produced in widths up to approximately 80 mm and with over 60 cores, they are extremely thin and, therefore, flexible (Fig. 7.8).

Multiplex systems

To the installation engineer, one of the advantages gained by the use of electronic equipment is that the amount of field wiring required is far less, in both quantity and size, than for the earlier power circuitry entailed by mechanical relay systems. Even further improvements are made possible by the use of multiplexing systems, i.e. the ability to convey a large number of signals each way along the same conductor, and these are, therefore, particularly suitable for installations requiring a large number of outstations, whether for data transmission or process control.

Since the early 1970s, optical fibre cables have been developed which

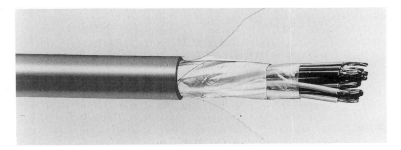

Fig. 7.7 BICC multicore cable for computer interfacing.

Fig. 7.8 Selection of BICC ribbon cables.

provide further advantages for light-current installations of all types; they have low attenuation and high bandwith, which reduces the necessity for repeaters, and are not subject to interference from heavy electrical equipment.

In hazardous areas, optical fibres give even greater safety than intrinsically safe circuits as the form of energy transmitted is, of course, lightwaves and not current (Fig. 7.9).

Fig. 7.9 BICC optical fibre TV camera cable.

Telephone cables

In the UK there is now greater freedom for the user to install his own internal telephone systems and, for this purpose, there is again a large range of cables available, all conforming with British Telecom specifications. The conventional type of multipair or multitriple cable consists of tinned copper conductors, PVC-insulated and sheathed with, in some cases, a nonmetallic rip-cord laid under the sheath to simplify stripping while, for under-the-carpet installations or situations where the conventional round cables are inconvenient or too bulky, ribbon cables are again available with up to 50 ways. Where such cables may be subject to damage or heavy traffic, such as under floor coverings, BICC produces ribbon cables insulated with cross-linked PVC (XLPVC) which is more robust than standard PVC. XLPVC should not be confused with XLPE-insulated cables which, among other advantages, have fire-retardant properties.

Cable jointing and terminating

Although the methods employed for jointing and terminating cables of all types have been simplified, largely due to the use of improved materials for insulation and sheathing, the importance of utilizing correct techniques and methods cannot be too strongly emphasized. All joints and terminations introduce potentially dangerous points; in power circuits a faulty joint will lead to local hot-spots with ultimate failure of the cable, while in light-current installations for process control, data transmission and communications a high resistance connection (dry joint) can be sufficient to prevent equipment from operating satisfactorily.

Multicore cables, whether for mains voltages or light-current duty, generally present the greater problem as the crutch, i.e. the point at which conductors are splayed out from the normal formation, constitutes a naturally weak area in which air may be trapped if a termination or joint is not correctly formed, leading to breakdown at a later stage. Consequently, operatives employed in jointing operations should preferably have attended, and been approved by, an authorized training centre, many of which are operated by the major cable manufacturers and regional electricity companies.

Single-core cables should, preferably, never be jointed, but where this is essential it should be effected only in purpose-made joint boxes equipped with suitable mechanical or compression-type connectors. These may be of the ferrule type with pinching screws or, as with terminations, bolted clamps requiring the bare conductor to be either wound around the bolt between shaped washers or enclosed in Ross—Courtney (crimped) type terminals which are then threaded over the screw thread and clamped. It is essential with all types of stranded cable to ensure that every strand is included in the joint or termination and, particularly with aluminium conductors, to follow cable manufacturers' recommendations for tightening torques. Aluminium, although lighter in weight and less expensive than copper, unfortunately has a higher coefficient of expansion and this has, at times, caused connections to slacken shortly after commissioning. It is, therefore, advisable for the installer of aluminium cables to recheck all clamp-type connections after electrical load has been applied. This does not imply, however, that a similar procedure is unnecessary with copper conductors but that it may not be so essential provided that connections are fully checked in the first instance.

Crimped terminals are quite adequate for the smaller, relatively lightly loaded cables but, otherwise, compression sleeves and lugs, provided that the recommended torques are applied, are unlikely to give rise to problems during the life of a cable under the most arduous circumstances.

Chapter 8
Cable Sizing

In any installation the importance of calculating and installing the correct size of cable for each circuit is paramount. This entails using the correct formula and correction factors given in the IEE Wiring Regulations. Voltage drop up to and in the final circuit must not be overlooked as this can have a bearing on the conductor size chosen.

It is therefore essential for the designer or installer to ascertain the conditions under which cables are to be installed, i.e., the external influences that are going to affect the cables, such as the ambient temperatures that will be encountered along the cable route or whether the cables will be in contact with thermal insulation. Contact with other cables along the route must not be ignored, since this can have an effect not only on the current-carrying capacity of the cables to be installed, but upon the cables they come into contact with. Further considerations which are the responsibility of the designer or installer are the type of protection to be installed (i.e., the type of fuse gear or circuit breakers to be used), the method of installation and the grouping of cables.

As far as voltage drop is concerned it is important to obtain accurate route lengths for the cables, together with the maximum continuous current the conductors will have to carry, to enable the voltage drop in each cable to be calculated.

It is important when considering voltage drop to remember that some electrical equipment, such as the direct-on-line start squirrel cage induction motor, can take a current some six to seven times the equipment's normal running current when it is started. This means that for a short period of time the voltage drop in the cables feeding the equipment may be higher than the 4% permitted by the IEE Wiring Regulations. An excessive reduction in voltage, even for a short period of time, may have an adverse effect on the starting characteristics of the equipment and in certain circumstances could affect its correct operation. Another important consideration concerning voltage drop is the effect that a sudden drop in voltage can have on electronic equipment. In these circumstances the operational characteristics dictate the allowable, as against the permitted, voltage variation.

Correction factors

The common term in use when considering sizing conductors with the 14th Edition of the Wiring Regulations was *derating factor* even though at ambient temperatures below 30°C an increased rating can result. The 16th Edition Wiring Regulations use the term *correction factor*, so both terms can be found in use.

The factors specifically referred to in the regulations are:

(1) C_g: grouping, more easily remembered as G, tables 4B1, 4B2 and 4B3.
(2) C_a: ambient temperature, more easily remembered as A, tables 4C1 and 4C2.
(3) C_i: thermal insulation, more easily remembered as T, table 52A.
(4) BS 3036 rewirable fuse, more easily remembered as S (see Regulation 433−02−03); no symbol is given in the regulations.

By using the symbols G, A, T and S a useful mnemonic GATS can be used to represent *Grouping*, *Ambient temperature*, *Thermal insulation*, *Semi-enclosed fuse*.

The first three items in the list affect the heat being dissipated from the conductor and are therefore thermal constraints. Item (4) is concerned with the characteristics of the BS 3036 fuse when it is being used as an overload device. How the above are used to size cables is indicated below.

Overloads

Before considering each of the items in the above list, consideration should be given to the reasons why derating factors have to be applied.

Where a protective device is being used for overload protection, the regulations call for the protective device to operate when the current reaches a value of 1.45 times the rating of the protective device. If the conductor is carrying the maximum permitted current, as given in the appropriate table from Appendix 4, the conductor operating temperature will also be at its maximum permitted value. Allowing the current to increase by 45% before the protective device has to trip the circuit will raise the conductor temperature (to approximately 114°C in the case of PVC insulated cables protected by HRC fuses or MCBs). This increase is acceptable since it will not damage the conductor's insulation, however, if the operating temperature of the conductor is higher than the permitted operating temperature for the conductor, the 45% increase in current due to an overload will raise the conductor's temperature to a level where the conductor's insulation would be damaged.

Grouping (G)

Where conductors are grouped together they interrelate with the heat dissipated from each other. This raises a conductor's operating temperature to a value that could exceed the permitted value. To ensure the operating temperature of the conductor will not exceed the permitted value, the effective current-carrying capacity of the conductor has to be reduced. This reduction is undertaken by using correction factors from Tables 4B1, 4B2 or 4B3 in the Wiring Regulations. The correction factor chosen is dependent upon how the conductors are installed and the number grouped together. The symbol used in the regulations for this factor is C_g (changed to G above).

Several formulae are given in the regulations for derating for grouping depending upon the conditions. These are summarized below, but since the formulae contain other symbols these are listed first.

Symbols

I_n Rating of the protective device
I_b Design or full load current of the circuit
I_t Calculated current-carrying capacity required for the conductor
I_{tab} Tabulated current-carrying capacity in the tables
t_f Final conductor operating temperature
t_p Maximum conductor temperature allowed by the current-carrying capacity tables
t_a Ambient temperature
C_t For operating temperature of conductor (Appendix 4 clause 7.1)

The last item in the list is the result of a calculation concerned with correcting the mV/A/m given in the tables for voltage drop, where the temperature of the conductor is not at the specified operating temperature for the conductor.

Circuits protected against overload

Ignoring BS 3036 fuses and socket outlet circuits for the moment, where circuits are being protected against overload and any number of circuits could be overloaded at the same time, the calculated value of current-carrying capacity I_t required for the conductors is given by:

$$I_t = \frac{I_n}{G} \qquad (8.1)$$

where I_n is the rating of the protective device in amperes and G is the correction factor from the appropriate table in Appendix 4.

Again ignoring BS 3036 fuses and socket outlet circuits, where circuits are being protected against overload, but not more than one circuit can have an overload condition occurring at any one time, i.e., simultaneous overload will not occur, two calculations are carried out as illustrated below:

First formula:

$$I_t = \frac{I_b}{G} \qquad (8.2)$$

Second formula

$$I_t = \sqrt{I_n^2 + 0.48I_b^2 \left(\frac{1 - G^2}{G^2}\right)} \qquad (8.3)$$

The conductor is then sized on the larger of the two calculated values of I_t from Equation (8.2) or (8.3).

Circuits not subject to overload

Where the circuit is not subject to overload as in the case of a fixed heating load the formula is changed to:

$$I_t = \frac{I_b}{G} \qquad (8.4)$$

where I_b is the design or full load current of the circuit. This formula is applicable to all types of protective device since the protective device is only providing fault current protection.

BS 3036 fuses

Where overload protection is carried out using BS 3036 fuses then the above equations are changed. In Equations (8.1) and (8.2) the result is divided by 0.725 and in Equation (8.3) the I_n^2 is multiplied by 1.9. This is due to the rewirable fuse having a fusing factor of 2 and not 1.45 as required by the regulations.

Socket outlet circuits

There is an exception to the above rules concerning socket-outlet circuits.

Due to the nature of these circuits the designer has no control over what load may be connected to the circuit. For radial circuits it must be assumed that the load on the circuit will be at least the rating of the protective device. When two or more circuits are grouped together it cannot be assumed that simultaneous overload will not occur and Equation (8.1) should be used.

Where the socket outlet circuit is a ring circuit, derating only takes place if more than two circuits are grouped together and then the appropriate correction factor from Appendix 4 is chosen such that the current-carrying capacity of the conductors is 0.67 times the rating of the protective device, i.e. for ring circuits:

$$I_t = \frac{0.67 \times I_n}{G} \tag{8.5}$$

If the ambient temperature exceeds 30°C, or the cables are enclosed in thermal insulation, the appropriate factor is placed in the denominator alongside the grouping factor G.

Lightly loaded conductors

If a conductor is expected to carry not more than 30% of its grouped rating, it can be ignored for the purposes of finding the correction factor for the rest of the grouped conductors.

This again requires two calculations, the first calculation sizes the lightly loaded cable by using the formula:

$$I_t = \frac{I_n}{G} \tag{8.6}$$

The lightly loaded cable is then sized to I_t and the tabulated value I_{tab} of current-carrying capacity of the conductor chosen is then noted.

The design current I_b and the tabulated current-carrying capacity I_{tab} for the lightly loaded cable from the appropriate table in Appendix 4 are then used in the following formula:

$$\text{Percentage of grouped rating} = \frac{I_b}{G \times I_{tab}} \times 100 \tag{8.7}$$

If the percentage is 30% or less, the conductor can be ignored when counting the remaining number of conductors grouped together.

Derating unnecessary

Where the spacing between conductors is at least twice the diameter of the largest conductor no derating need be applied.

Ambient temperature (A)

As can be seen from the tables in Appendix 4 of the regulations, the permitted operating temperature of the conductor varies according to the type of conductor insulation for example, 70°C for PVC, 90°C for thermosetting insulation (XLPE) and 85°C for rubber.

Additionally, the current-carrying capacities given in the tables in Appendix 4 are based on an ambient temperature of 30°C, so a higher ambient temperature will reduce the rate of flow of heat out of the conductor, raising the conductor's operating temperature above the value permitted. This means that the current-carrying capacity of the conductor has to be reduced to compensate for the reduction in the heat lost from the conductor.

Two tables of correction factors are provided, 4Cl and 4C2, the latter covers BS 3036 fuses and the former all other types of protective device. The correction factor has to be applied irrespective of the distance the conductors are installed in a higher ambient temperature area and as such, designers should avoid installing cables in these areas as much as possible. It should be noted that the ambient temperature is the temperature of the surrounding media and does not include the temperature of the equipment.

Correction for ambient temperature is carried out using the following formula:

$$I_t = \frac{I_n}{A} \tag{8.8}$$

Thermal insulation (T)

Where a cable is installed in an area to which thermal insulation is likely to be applied, the cable should be installed in such a position that it will not be covered by insulation. Where this is not possible the cross-sectional area of the cable has to be increased.

Where the cable is in contact with insulation on one side the other side being in contact with a thermally conducting surface, the appropriate table in Appendix 4 should be used for choosing the cable with the correct current-carrying capacity. Where the cable is totally enclosed in thermal insulation for lengths exceeding 0.5 m the derating factor is 0.5, where the length is less than 0.5 m the derating factor varies with the length the cable is in contact with thermal insulation and is given in Table 52A in the IEE Wiring Regulations.

The formula for calculating the current-carrying capacity required for the cable is as follows:

$$I_t = \frac{I_n}{T} \tag{8.9}$$

General

I_n has been used in the above formulae, but where the conductor is not liable to be overloaded and the protective device is only providing fault-current protection I_b should be used in place of I_n.

Voltage drop

The effects of voltage reductions below the nominal standard, range from a mere lowering of the efficiency to a complete inability of plant and equipment to function at all: consequently, it is essential to ensure that any drop in voltage is kept to a minimum.

The regulations now call for the voltage at the terminals of any fixed current-using equipment to be greater than the lower limit given in the British Standard for the relevant equipment or, where no such standard exists, the voltage at the terminals of the equipment shall be such that the safe functioning of the equipment is not impaired. However, these requirements are deemed to be satisfied if the voltage drop from the origin of the installation to the end of the final circuit does not exceed 4% of the declared nominal voltage, providing the supply is made in accordance with the Electricity Supply Regulations 1988 (amended). This is a major change from the 2.5% voltage drop allowed for decades by the regulations. As mentioned earlier the operational characteristics may dictate the voltage drop allowed.

The voltage drop is given in the regulations in the form of mV/A/metre, based on the conductor working at the operational temperature given in the current-carrying capacity tables. The voltage drop formula is therefore:

$$V_D = I_b \times L \times mV/A/m_z \times 10^{-3} \tag{8.10}$$

where I_b is the full load current of the circuit and L is the circuit length.

Adjustment for temperature

In practice the conductors will rarely be loaded to the full current-carrying capacity given in the tables, so an adjustment can be made to the mV/A/m for the lower conductor temperature.

Where HRC and MCB protective devices are used and the ambient temperature is not less than 30°C, the following formula will give the temperature of the conductor when it is not carrying its tabulated current-carrying capacity.

$$t_f = t_p - \left(G^2 A^2 - \frac{I_b^2}{I_{tab}^2}\right)(t_p - t_a) \tag{8.11}$$

The correction factor for operating temperature then becomes:

$$C_t = \frac{230 + t_f}{230 + t_p} \tag{8.12}$$

This formula can of course be related to the required mV/A/m and the actual mV/A/m given in the tables.

For conductors up to 16 mm², C_t is applied to the total mV/A/m, but for conductors larger than 16 mm², C_t is only applied to the resistance portion of the mV/A/m, since temperature has no effect upon reactance. The voltage drop formula for conductors up to 16 mm² becomes:

$$V_D = C_t \, mV/A/m_z \times I_b \times L \times 10^{-3} \tag{8.13}$$

and for conductors larger than 16 mm² the formula becomes:

$$V_D = I_b \times L \times \sqrt{C_t \, mV/A/m_r^2 + mV/A/m_x^2} \times 10^{-3} \tag{8.14}$$

mV/A/m$_r$ and mV/A/m$_x$ are the resistive and reactive components, respectively, of mV/A/m$_z$ from the Tables in Appendix 4 of the Regulations.

Adjustment for power factor

As far as the industrial installation is concerned, where the load has a power factor, part of the current flowing down the conductor is reactive, so an adjustment can also be made for the power factor of the circuit.

The reactive component of the current can be ignored for conductors up to 16 mm² therefore the total mV/A/m from the tables is multiplied by the power factor cos φ.

The formula for voltage drop for conductors up to 16 mm² becoming:

$$V_D = \cos \phi \, mV/A/m_Z \times I_b \times L \times 10^{-3} \tag{8.15}$$

For conductors 25 mm² and above the formula for voltage drop becomes:

$$V_D = (\cos \phi \, mV/A/m_r + \sin \phi \, mV/A/m_x) \times I_b \times L \times 10^{-3} \tag{8.16}$$

Adjustment for both temperature and power factor

The above formulae can be combined so that adjustment for both temperature and power factor can be carried out at the same time. This is achieved by multiplying the resistive part of the formula by C_t, i.e., the formula now becomes:

For conductors up to $16\,mm^2$

$$V_D = C_t \cos \phi \ mV/A/m_z \times I_b \times L \times 10^{-3} \tag{8.17}$$

and for conductors $25\,mm^2$ and above

$$V_D = (C_t \cos \phi \ mV/A/m_r + \sin \phi \ mV/A/m_x) \times I_b \times L \times 10^{-3} \tag{8.18}$$

Now $\cos \phi$ is the power factor for the circuit and $\sin \phi$ can be found from tables or by using the formula:

$$\sin \phi = \sqrt{1 - \cos^2 \phi}$$

Additional considerations when sizing conductors

Carrying out a calculation to ensure that the conductor's operating temperature does not exceed the permitted amount, or that the voltage drop in the circuit is satisfactory, are not the only calculations that have to be considered.

Consideration has to be given to the thermal and mechanical effects upon conductors caused by short-circuit currents and earth fault currents. In addition, consideration has to be given to the phase earth loop impedance so that the circuit will be disconnected within a specified time with a phase to earth fault.

Short-circuit current

Calculations may be necessary to ensure that the circuit conductors are protected against short-circuit current and that equipment is capable of withstanding any peak short-circuit current that is likely to flow through it.

As far as conductors are concerned the formula to use is:

$$I^2t = k^2S^2$$

where I is the prospective short-circuit current in amperes, t is the time taken for the protective device to disconnect the circuit, S is the cross-sectional area of the conductor in m^2 and k is a factor based on the type of conductor material, the conductor temperature at the start of the fault and the limit temperature of the conductor's insulation; k represents the maximum thermal capacity of the conductor for the type of insulation being used. A typical value of k for a copper conductor having PVC insulation and operating at a temperature of 70°C is 115.

Where the disconnection time of the protective device is less than 0.1 s then I^{2}t is the energy let through the protective device as specified by the manufacturers. To ensure that the conductor is protected by the protective device I^{2}t must never exceed k^{2}S^2.

Protection against indirect contact

Circuits have to be checked to ensure that they will disconnect within a specified time when an earth fault occurs. The disconnection time for the circuit depends upon the voltage to earth and whether the equipment is fixed or used for hand held equipment.

The calculation entails working out the phase earth loop impedance for the circuit. This is illustrated in Fig. 8.1.

As can be seen from Fig. 8.1, Z_A, Z_B, Z_C and Z_D are external to the installation and are referred to as Z_E (external earth loop impedance). As far as large installations are concerned it is necessary to consider Z_E as being external to the circuit under consideration. The total of Z_A, Z_B, Z_C, Z_D, Z_1 and Z_2 is the total earth loop impedance and is referred to as the system impedance Z_S.

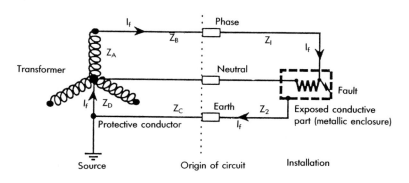

Phase earth loop = $Z_A + Z_B + Z_1 + Z_2 + Z_C + Z_D$

Fig. 8.1 Phase earth loop impedance. (Taken from the *Handbook on the IEE Wiring Regulations* by T.E. Marks.)

Calculations

The resistance of the conductor is taken at the average of the permitted operating temperature of the conductor and the limit temperature for the conductor's insulation. Where the conductor exceeds 35 mm² the impedance of the conductor is taken instead of its resistance.

Where the protective conductor is installed in the same cable as the phase conductor its operating temperature is taken as being the same as the phase conductor, but where it is not in contact with other cables its operating temperature is taken to be ambient temperature, which in the UK is taken as 30°C.

For example, a twin and CPC cable having PVC-insulated copper conductors, the maximum permitted operating temperature is 70°C and the limit temperature for the PVC insulation is 160°C. The average is therefore 115°C and this is the temperature that would be used to determine the conductor's resistance.

Where the protective conductor is steel conduit, trunking or the steel wire armouring of cables, the impedance of the protective conductor is taken, not its resistance.

Protective conductors

The phase earth loop impedance of the circuit can also be used in checking whether the circuit's protective conductor has sufficient thermal capacity to withstand any fault current that is likely to flow through it.

The formula for checking that the protective device is suitable is derived from the formula for short-circuit current.

$$S = \frac{\sqrt{I_f^2\, t}}{k} \tag{8.19}$$

where I_f is the fault current, t is the time taken for the protective device to disconnect the fault current and k is a factor obtained from the tables in Chapter 54 of the IEE Wiring Regulations. It should be noted that the value of k is not inside the square root as shown in many textbooks. For a PVC-insulated protective conductor in contact with other current carrying conductors the value of k would be 115 for a copper conductor or, where it is installed independently of other conductors, k would be 143.

The fault current I_f is determined by dividing the phase to earth voltage U_O by Z_S. This fault current is then used to determine the disconnection time of the protective device. This is illustrated in Fig. 8.2 where the protective device is a BS 88 fuse, I_f is given as 120 A and the resultant

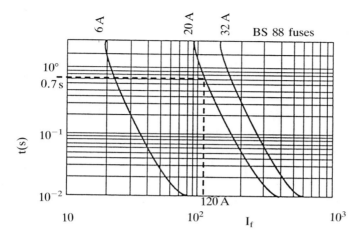

Fig. 8.2 Disconnection time. (taken from the *Handbook on the IEE Wiring Regulations* by T.E. Marks.)

disconnection time is 0.7 s. All that is then required is to choose the correct k factor from the appropriate table in Chapter 54.

There is an alternative for checking that the protective conductor is suitably sized and that is to use Table 54G. This table will, in general, stipulate a conductor size far in excess of that actually required. In any event the earth loop impedance for the circuit still has to be worked out to check the disconnection times for indirect contact. Table 54G must not, however, be used if the fault current is expected to be less than the short-circuit current (Regulation 543−01−01)

Other considerations

The size of conductors is not solely determined by the parameters discussed above, consideration has to be given to the heating effect placed on conductors when equipment that has an inrush current is frequently started and stopped. This is equivalent to having a permanent overload on the conductors and its connections.

Examples of sizing conductors

Example 1

Four circuits are to be wired in PVC cable enclosed in the same conduit and trunking, the supply to each circuit being from a 40 A HRC fuse. If the ambient temperature is 30°C and the circuits are not in contact with thermal insulation, what size cables will be required:

(a) If the circuits are feeding 3-phase 3-wire BS 4343 socket outlets.
(b) If the circuits are feeding single-phase instantaneous shower heaters with a full load current of 36 A?

Working (a)

The derating factor for ambient temperature A and thermal insulation T will be 1. From Table 4B1 Appendix 4, for four circuits enclosed in trunking the derating factor is 0.65.

The circuits are feeding socket outlets from a 40 A fuse so simultaneous overloading could occur, therefore Equation (8.1) has to be used.

$$I_t = \frac{I_n}{G} = \frac{40\,A}{0.65} = 61.54\,A$$

From column 5 of Table 4D1A 16 mm^2 cable is required with an I_{tab} of 68 A.

Working (b)

Full load current for the circuit is 36 A. Derating factors A and T will equal 1, derating factor for grouping = 0.65.

The loads are resistive and fixed so it is assumed that simultaneous overloading will be unlikely, therefore Equations (8.2) and (8.3) can be used.

Equation (8.2)

$$I_{t1} = \frac{I_b}{G} = \frac{36}{0.65} = 55.38\,A$$

Equation (8.3)

$$I_{t2} = \sqrt{I_n^2 + 0.481_b^2\left(\frac{1 - G^2}{G^2}\right)} = \sqrt{40^2 + 0.48 \times 36^2\left(\frac{1 - 0.65^2}{0.65^2}\right)}$$

$$I_{t2} = \sqrt{40^2 + 0.48 \times 36^2 \times 1.37} = 49.52\,A$$

The conductor must be sized on the larger of I_{t1} and I_{t2}. In this instance the larger value 55.38 A and from column 4 of Table 4D1A 10 mm^2 cable would be required. If the circuit had also been affected by a high ambient temperature, and/or had been enclosed in thermal insulation, the 55.38 A would then have been divided by the appropriate derating factor for ambient temperature, and/or thermal insulation, and the revised I_t would then be used to size the conductor.

Comments on working

Circuits with fixed resistive loads are unlikely to go into overload. The circuit protective device is therefore only providing fault current protection. The derating of the conductors is only carried out to ensure the conductor's normal operating temperature does not exceed the value permitted by the tables. Consequently, Equations (8.2) and (8.3) tend to make Equation (8.4) redundant. Equation (8.4) would be used for fixed resistive loads when those circuits were mixed with circuits that could be overloaded and simultaneous overloading could occur for instance, if the shower circuits were mixed with the socket-outlet circuits mentioned above. In both (a) and (b) the conductor size required is probably too large to fit in the equipment terminals, so the grouping would have to be reduced.

Example 2

What size cable would be required in Example 1 if the socket outlets were:

(a) Protected by a BS 3036 fuse and
(b) Part of the trunking run is through an ambient temperature of 45°C and part of the trunking run passes through a wall with 500 mm of insulation?

Working 2(a)

The formula now becomes:

$$I_t = \frac{I_n}{G \times S} = \frac{40}{0.65 \times 0.725} = 84.88 \, A$$

The cable size required from Table 4D1A column 5, 25 mm^2 cable required.

Working 2(b)

The installation now looks like that given in Fig. 8.3. The correction factors will be:

(1) Grouping 0.65.
(2) Ambient temperature is 0.91 from Table 4C2 (semi-enclosed fuse).
(3) Thermal insulation 0.5 from Regulation 523−04−01.
(4) Semi-enclosed fuse 0.725 from 433−02−03.

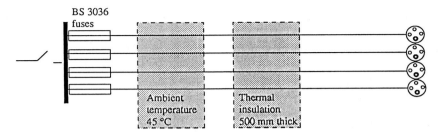

Fig. 8.3 Circuit for problem 2(b).

This example uses all the derating factors. With the exception of the semi-enclosed fuse they are not all applicable at the same point in the cable run. This means that they are only applied to that portion of the cable run which they affect. If the cable run is reasonably short the cable can be sized to the factor that will give the largest conductor size.

In the above case the worse factor is for the section which is grouped in thermal insulation, so any conductor that satisfies this section will automatically satisfy the others. The formula now becomes:

$$I_t = \frac{40\,A}{0.65 \times 0.5 \times 0.725} = 169.76\,A$$

From column 5 of Table 4D1A the size of cable required is $70\,mm^2$.

Note that the derating factor for the semi-enclosed fuse is still applicable since it affects the whole circuit.

Example 3 Ring circuits

Ten ring circuits are to be installed in trunking round an office. The ambient temperature is normal and the trunking is not in contact with thermal insulation. Calculate the size of cable to be installed if:

(a) The protective device is a 32 A MCB.
(b) The only items the socket outlets are supplying are typewriters and desk-top computers with a total connected load of 9.8 A per circuit.

Working 3(a)

The designer has no control over what load may be connected to the ring by the user. This is particularly true when the central heating fails and electric heaters are brought in to enable work to continue.

This means that simultaneous overload could occur and the correct formula to use is Equation (8.5) from above.

Working 3(a)

Ten circuits grouped in trunking will have a derating factor of 0.48 from Table 4B1.

$$I_t = \frac{0.67 \times I_n}{G \times A \times T} = \frac{0.67 \times 32\,A}{0.48 \times 1 \times 1} = 44.67\,A$$

The size of cable required would be $10\,mm^2$ from Table 4D1A, which would not fit in the socket outlets, the number of circuits grouped together must therefore be reduced.

Working 3(b)

This is a case where the designer limits what the client can use on the ring circuit. If the total load on each circuit is only 9.8 A the circuit can be protected by a 10 A protective device. The distribution board must, however, be marked to show that this is the maximum permitted size of protective device for these circuits. The client should also be advised as to the maximum load allowed on each circuit.

The circuits can still be overloaded i.e., plugging in a 3 kW heater (12.5 A) would not trip the protective device thus overloading the circuit, so the same formula as 3(a) is used.

$$I_t = \frac{0.67 \times 10\,A}{0.48} = 13.95\,A$$

This clearly shows that using $1.5\,mm^2$ cable for the ring circuit would be all right for grouping, although the designer may want to build in a safety factor and use $2.5\,mm^2$ cable.

Example 4 Lightly loaded conductors

A trunking is to contain 16, 1.2 A control circuits each being protected by a 2 A fuse. The trunking also contains four motor circuits with a full load current of 38 A and protected by 40 A fuses. Determine the size of cable required.

Working (4)

If the conductor is not expected to carry more than 30% of its *grouped rating* it can be ignored when counting the number of circuits grouped together.

The first course of action is to determine the conductor size for the lightly loaded cables. The total number of circuits grouped together $= 20$; derating factor is 0.38, Table 4B1

$$I_t \text{ for control circuits } = \frac{I_n}{G} = \frac{2\,A}{0.38} = 5.26\,A$$

From Table 4D1A column 4 the minimum size of conductor for the control circuits is $1.0\,mm^2$ with an I_{tab} of $13.5\,A$.

Now check to see whether these circuits can be ignored when sizing the other conductors.

$$\% \text{ of grouped rating } = \frac{I_b}{G \times I_{tab}} \times 100 = \frac{1.2\,A}{0.38 \times 13.5\,A} \times 100$$
$$= 23.39\%$$

This percentage does not exceed 30% so the control circuits can be ignored when sizing the motor circuits.

Four motor circuits will have a grouping factor of 0.65 from Table 4B1

$$I_t = \frac{I_n}{G} = \frac{40\,A}{0.65} = 61.54\,A$$

These circuits would require $16\,mm^2$ cable from Table 4D1A. Without this relaxation the size required would have been $35\,mm^2$.

An alternative way of working out this problem is to use the I_{tab} for the cable selected in the following formula: $I_{tab} \times G \times 0.3 = I_b$ (maximum design current allowed for the lightly loaded cable). Which in the above example would have been $1.54\,A$.

Example 5

If four single-phase circuits each use $10\,mm^2$ cable with an I_{tab} of $57\,A$ to carry a load of $34\,A$, calculate the voltage drop in the cable if the route length is $60\,m$ and the circuits are installed in trunking:

(a) Using the mV/A/m given in the regulations.
(b) Taking into consideration the actual operating temperature of the conductor.

Working 5 (a)

From Table 4D1B column 3 the mV/A/m for $10\,mm^2$ cable is $4.4\,mV/A/m$

$$V_D = 34\,A \times 60\,m \times 4.4 \times 10^{-3} = 8.98\,V$$

Working 5 (b)

The actual operating temperature of the cables can be calculated from Equation (8.11)

$$t_f = t_p - \left(G^2A^2 - \frac{I_b^2}{I_{tab}^2}\right)(t_p - t_a) \qquad (8.11)$$

$$t_f = 70 - \left(0.65^2 - \frac{34^2}{57^2}\right)(70 - 30) = 67.33°C$$

The actual mV/A/m for the cables will be the ratio of the conductor's temperature applied to the mV/A/m given in the regulations. This can be achieved by using Equation (8.12)

$$\frac{\text{actual mV/A/m}}{\text{regulations mV/A/m}} = \frac{230 + t_f}{230 + t_p} \qquad (8.12)$$

therefore actual mV/A/m = regulations mV/A/m $\times \dfrac{230 + t_f}{230 + t_p}$

Voltage drop $V_D = 34\,A \times 60\,m \times 4.4 \times \dfrac{230 + 67.33°C}{230 + 70°C} \times 10^{-3}$

$$= 8.896\,V$$

The examples above show how to proceed in sizing cables. They are not comprehensive, since there is insufficient space available in a book of this type, but examples are given of short-circuit protection in Chapter 9 and protection against indirect contact in Chapter 12.

Chapter 9
Protective Devices

Low voltage systems

In the majority of low voltage installations the protective devices described in this chapter are adequate for the required duty, as these do not have to deal with excessively high prospective fault currents even if the installation is close to the transformer. In the case of a consumer accepting supply at high voltage the fault current for a 500 kVA transformer is unlikely to exceed 15 kA at the low voltage terminals while that of a 1000 kVA is only double that value for a phase-to-earth fault. On a three-phase circuit the comparable figures for a phase-to-phase fault (short-circuit) are 12 kA and 24 kA. These may vary according to the impedances of the supply system but the regional electricity companies, for single-phase circuits, have standardized on the figure of 16 kA, the others being proportionately higher. It is possible, of course, for transformers to be operated in parallel, in which case the above figures are much higher unless they are a considerable distance apart, as is usual with the regional electricity companies. In some cases, however, the paralleling of transformers is prevented by interlocking to keep down the short-circuit levels.

Fuses

There are four basic types of low voltage fuse used in consumers' installations, these being the semi-enclosed (rewirable) fuse, cartridge fuses to British Standard BS 1361 and high rupturing capacity (HRC) fuses to BS 88 Parts 2 and 6, with a miniature cartridge fuse to BS 1362. The semi-enclosed fuse has been widely used in the past for domestic installations and to some extent in industry and commerce but, although it is still recognized by the IEE Regulations, is not to be recommended for other than domestic installations. The disadvantages of the semi-enclosed fuse are that it is subject to deterioration with time; it has a higher fusing factor than cartridge fuses; the time/current characteristics are unstable making discrimination difficult; and it only has a low short-circuit breaking capacity.

With regard to the fusing factor, the cable rating tables in the IEE

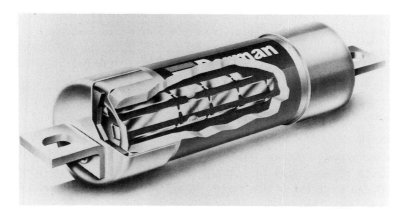

Fig. 9.1 Dorman Smith BS 88 cartridge fuse construction.

Wiring Regulations are based on cartridge fuses and, therefore, cables installed in conjunction with semi-enclosed fuses require to be de-rated by a factor of 0.725 when used for overload protection, this being in addition to the other correction factors explained in Chapter 8. This, on large installations particularly, leads to increased cable costs as, for a given size of semi-enclosed fuse, installed cables are larger than with cartridge fuses.

Both BS 1361 and BS 88: Parts 2 and 6 (Fig. 9.1) relate to fuses of the cartridge type, the main difference being that the first is limited to a maximum rating of 100 A and, therefore, is more applicable to smaller installations. They differ in appearance also, in that BS 88 fuses are fitted with fixing lugs while the smaller rated ones simply clamp into the fuse-carrier. The operating characteristics of each are very similar, as may be seen in Appendix 3 of the IEE Wiring Regulations.

As the fusing factor for cartridge fuses is 1.45 and the short-circuit rating high (BS 1361, 16.5 kA; BS 88, 80 kA), they are suitable for use on the types of installation under consideration.

The smaller cartridge fuses, to BS 1362, are available in 2, 3, 5, 10 and 13 A sizes and are intended for use in the flat-pinned 13 A BS 1363 plug and fused connection units as the immediate protection for equipment.

Care must be taken in the design of an installation to ensure that discrimination is obtained between series protective devices and, in this respect, the nondeteriorating characteristics of cartridge fuses offer advantages over the semi-enclosed type. (See the section on protective discrimination below.)

Although the cartridge fuse is a seemingly simple piece of equipment, its effectiveness is only obtained by the application of a high degree of development technology. This is clearly apparent from the range of fuses for each different application manufactured by the major producers such as GEC Alsthom Installation Equipment Ltd, a selection of which are shown in Fig. 9.2.

Miniature circuit-breakers

Although the different types of fuse have given, and still are giving, excellent service for many years, miniature circuit-breakers (MCBs) have increased in popularity and improved in design since they were first introduced to an extent that makes them more suitable in many ways for loads of up to 100 A (their maximum rating under BS 3871 Part 1) at 415 V (Fig. 9.3).

Added to their technical benefits are the advantages that it is possible to check the operation of an MCB after installation and, subject to the manufacturers' approval, to use it for normal switching operations, neither of which is applicable to fuses.

The MCB is a factory-moulded, sealed unit available with different types of terminal so that it can be simply plugged in to or bolted direct to

Fig. 9.2 Selection of HRC fuses including BS 88, BS 1361 and BS 1362 (plugtop) types. (Courtesy of GEC Alsthom Installation Equipment Ltd.)

Fig. 9.3 25 A 415 V type M9 MCB with padlock facility. (Courtesy of Merlin Gerin.)

busbars in a suitable housing, or used as a single-phase or linked three-phase unit. Protective features include magnetic-hydraulic, thermal-magnetic and assisted bimetal tripping mechanisms providing time-delay characteristics and suitability for overload and short-circuit protection. The residual current circuit-breaker (RCCB) referred to elsewhere is a further development of the MCB, providing protection against low earth fault currents also (Fig. 9.4).

As in the case of cartridge fuses, it is possible to use smaller cables with MCBs than with semi-enclosed fuses (rating for rating) when protecting

Fig. 9.4 Combined MCB/RCD comprising 16 A type 2 M9 MCB with 30 mA RCD. (Courtesy of Merlin Gerin.)

against overload current as the fusing factor is 1.45 against the 2.0 of the latter.

The major use of MCBs in commercial and industrial installations is for the protection of final circuits because of their relatively low short-circuit capacity, although models are available with ratings up to 16 kA, the majority only have a maximum breaking capacity of 9 kA. As most domestic installations, however, have much lower prospective fault levels at the origin of supply, MCBs are frequently installed in consumer units.

Moulded case circuit-breakers

Where short-circuit levels are high, moulded case circuit-breakers (MCCBs) of similar construction to MCBs are available. Maximum ratings differ between manufacturers, some being in excess of 3000 A, and with most makes it is possible to obtain numerous add-on facilities such as shunt release, undervoltage trips, current limiters and auxiliary contacts. For installation they are suitable as free-standing units or for building into compact cubicle-type switchboards (Fig. 9.5).

As would be expected, fault-current ratings for MCCBs are also much higher than for MCBs, which permits their use in closer proximity to the source of supply where, normally, fault levels are also higher; figures quoted by one major British manufacturer are 22 kA for the basic unit or 46 kA with current-limiters. For reference, the lower figure approximates to the prospective fault-current at the low voltage terminals of a 1 MVA 415 V transformer (ignoring any high voltage impedance).

Therefore, as the cable between the transformer and the MCCB increases the total impedance, thus reducing the fault-current even further, such a unit is, in many cases, quite adequate for installation on the main switchboard. The slightly higher cost of current limiters, however, may be justified in the event of a low power factor fault as this increases the amount of energy stored in a system; the energy is dissipated through the protective device under fault conditions and the MCCB must be capable of dealing with this.

Typical dimensions of a three-phase 415 V MCCB in the 600 to 800 A range are 560 mm high, 230 mm wide by 150 mm deep, measurements which give a considerable space-saving advantage over a comparable fuse-switch being approximately half the size of the latter (Fig. 9.6).

Although an MCCB is relatively maintenance free, it must be stressed that there are limits to the life and performance capability which are available from the manufacturers and must be considered for each application. MCCBs are covered in BS EN 60947−2.

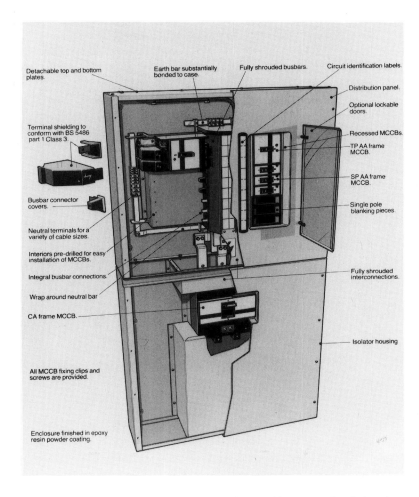

Fig. 9.5 Dorman Smith 630 A 'Loadbank' moulded case circuit-breaker, sectioned distribution panel.

Air circuit-breakers

One of the oldest forms of automatic protective device is the air circuit-breaker (ACB), many of the earlier patterns still being in service at the present time. The ACB is currently covered by BS EN 60947−2: Part 1, which also includes the MCCB, effectively the modern replacement for use in the range of commercial and industrial installations covered in this book.

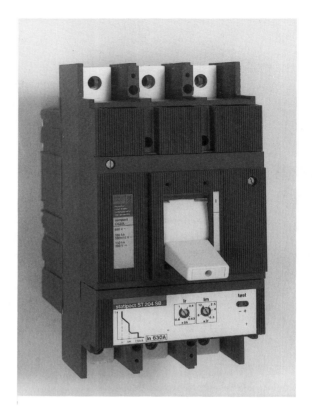

Fig. 9.6 Typical moulded case circuit-breaker.

Air is, obviously, the cheapest form of insulant but has the disadvantage that it contains dust and moisture, and these could lead to breakdown of a gap due to ionization. In the air circuit-breaker, an arc-chute lengthens the arc as the ACB contacts open and, therefore, a greater arc voltage is required to sustain it than is available from the supply. Arc-chutes take many forms but all perform the same function of lengthening the arc. The form of construction of the ACB reduces dust and moisture problems while in operation, but care must be taken when a unit is opened up for maintenance purposes to ensure that foreign bodies are completely excluded. This difficulty is removed by the use of the MCCB which, besides being a sealed unit, is usually cheaper than an ACB, although the more sophisticated designs with extensive adjustable protection arrangements reduce, or eliminate entirely, the cost advantage of an MCCB.

Various ancillary devices are available for ACBs which provide all the necessary protection against overcurrents and earth faults, with or without time delay features, either direct acting or by means of relay systems. Usually, facilities are provided for easy readjustment on site to meet changed conditions.

Unlike the MCCB, this type of equipment must receive regular inspection and maintenance (depending upon the severity of duty and frequency of operation) as, even with arc control, the main and secondary contacts deteriorate through burning and oxidization.

High voltage systems

As the operational requirements of high voltage systems are far more onerous than low voltage systems, the range of protective devices is on a much wider scale and, to some extent, manufacturing costs are not as important as performance. Consequently, developments have taken place in circuit-breakers, particularly since the early 1960s, which, although equally suitable for low voltage systems, make them uneconomical for this purpose. Despite this, several types of high voltage circuit-breaker have their counterparts for low voltage duties although the latter, due to the lower insulation levels and operational performance required, are much smaller.

Fuses — high voltage

Two main types of high voltage fuse are available in the UK: the liquid-filled and the HRC cartridge-type.

The liquid-filled fuse, which has been more commonly used by the supply industry for rural high voltage systems, consists of a glass tube in which the fuse element is attached to the sealing cap at one end and to a spring-loaded piston at the other, the end of the spring being connected to the other sealing cap. Before sealing, the tube is filled with an insulating and fire-extinguishing liquid.

The operation of the fuse releases the spring which retracts, drawing the piston up the tube and forcing the liquid down over the arc to extinguish it rapidly.

The disadvantages of the liquid-filled fuse are that, by the nature of its construction, it requires very careful handling. This type is now being phased out by the HRC fuse for all voltages.

For general use on systems up to 11 kV the cartridge fuse, which is of similar construction to the low voltage type, has been used for many years as it meets most, if not all, of the requirements of the smaller installation.

These fuses are available with or without striker pins which, in similar fashion to the piston in the liquid-filled fuse, are released when a fuse blows and ejected from one end of it. The force of the striker pin is sufficient to enable it to operate tripping mechanisms in equipment such as automatic fuse-switches or simply to provide more positive indication that a fuse has operated. A number of the GEC Alsthom fuses shown in Figs 9.7(a) and (b) have this facility.

Oil circuit-breakers

For high voltage switching and protection on commercial and industrial

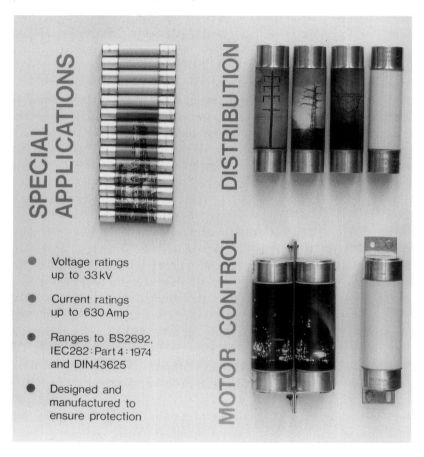

Fig. 9.7(a) Selection of HRC fuses for voltages up to 34.5 kV indicates the numerous applications and requirements. (GEC Alsthom Installation Equipment Ltd.)

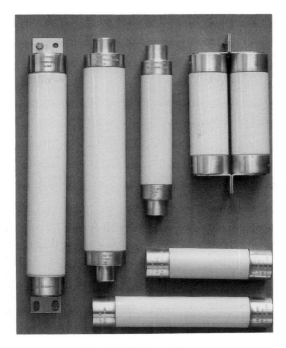

Fig. 9.7(b) (contd).

installations the oil circuit-breaker (OCB) is still popular, although the recently introduced SF$_6$ and vacuum circuit-breakers eliminate the fire risk associated with the OCB.

There are two basic designs of OCB for use on systems up to 11 kV: single-break and double-break. In both cases the contacts are enclosed in an arc control device, the whole contact system being contained in an oil-filled tank. Bushings fixed to the contact system protrude from the top of the circuit-breaker tank to provide means of connection to the incoming and outgoing circuits.

The circuit-breaker, which can be of either fixed or withdrawable design, is contained within a metal-clad cubicle. A withdrawable OCB is mounted on a 'hospital' truck which can be isolated from the cubicle by first disconnecting the breaker from the live circuits (usually by a racking mechanism) and then wheeling it out of its cubicle. As this operation is performed, shutters inside the cubicle automatically close over the live orifices to prevent accidental contact. Extensive interlocking prevents incorrect operation. Fixed OCBs need to be provided with isolating devices, usually on both sides of the unit, i.e. on both busbar and cable sides.

Fixed and withdrawable circuit-breakers are available as single units with cable boxes or with busbars, allowing several units to be assembled as a composite switchboard for the larger installations.

Assembly of this type of switchgear on site, although not necessarily complicated, must be accurately effected. Foundations must be adequate and true, fixing bolt-holes suitably located, etc. and, although all manufacturers will supply details and foundation drawings, it is advisable to employ them for the assembly as they have their own expert teams for such work and, possibly, also for commissioning (Fig. 9.8).

Vacuum circuit-breakers

In appearance the vacuum circuit-breaker (VCB) is similar to the cubicle-type OCB although it is more compact due to the omission of the oil tank. In the VCB, the fixed and moving contacts are factory-assembled in ceramic 'pots' or 'bottles', each phase being a separate unit which, after the air has been evacuated, is sealed-for-life; the vacuum interrupters are then bolted to terminals on the rear of the operating mechanism. The VCB is a very reliable breaker with high performance characteristics and low maintenance requirements. Although the individual bottles are not

Fig. 9.8 Yorkshire Switchgear Ltd 'SO-HI' indoor metal-clad oil circuit-breaker with moving portion withdrawn − suitable for voltages up to 15 kV and 1250/2000 A.

repairable (the main cause of failure being loss of vacuum), it is relatively simple to change them in the event of failure. However, VCBs have been in operation since the 1960s with very few failures to date.

The advantages of this type of equipment over the OCB make it suitable for commercial and medium-sized industrial installations, particularly when maintenance expertise is at a minimum (Fig. 9.9).

Sulphur hexafluoride circuit-breakers

This type of circuit-breaker, in a similar manner to the VCB, also excludes the use of oil, the contacts being in a sealed enclosure containing sulphur

Fig. 9.9 A 15 kV Reyrolle double tier type YMV vacuum circuit-breaker showing one circuit-breaker in the maintenance position.

hexafluoride (SF_6) gas which has good arc extinguishing properties and high dielectric strength. As these breakers also have few maintenance requirements, they too are suitable for the type of installation considered although, to date, their application has been largely confined to extensive installations and the power supply industry (Fig. 9.10).

Minimum oil circuit-breakers

The OCBs previously referred to are equipped with comparatively large oil tanks containing many litres of oil, with other insulation confined to the current carrying parts. As implied by the name, the minimum oil circuit breaker has a low oil content used mainly for the quenching of arcs. As this reduces the dielectric strength to a considerable extent, further solid insulation is included on the inside faces of the circuit breaker tank.

Fig. 9.10 Yorkshire-Switchgear Ltd YSF6 circuit-breaker suitable for voltages up to 24 kV, 31.5 kA.

The minimum oil circuit-breaker is of continental manufacture and has not gained much popularity in the UK.

Ring main units

The high voltage ring main distribution system described in Chapter 4 generally employs package ring main units installed at the required positions rather than by conventional switchboards as they are relatively simple to install and economically priced. The basic industrial ring main unit consists of a central circuit-breaker with a load-break isolator on either side. It is usual for the switchgear to be oil filled and the complete unit is suitable for indoor or outdoor installation (Fig. 9.11).

Alternative methods of employing the ring main unit are shown in Figs 4.1 and 4.2. Figure 9.12 is the more recently developed SF_6 unit.

Oil switches and switch-fuses

When operational requirements are anticipated to be minimal as, for example, on a ring main unit for simple load-transfer operation, the oil switch is the most economical device and is quite suitable for the duty. It is manually operated, the handle being connected by an arrangement of

Fig. 9.11 Yorkshire Switchgear Ltd TYKE 13.8 kV, 21.9 kA oil-filled ring main unit.

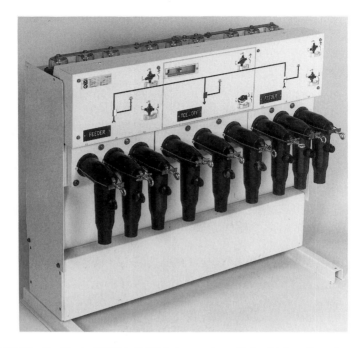

Fig. 9.12 Merlin Gerin RM6 'all SF$_6$' ring main unit for high voltage/low voltage transformer substation with SF$_6$ transformer protection circuit-breaker.

springs and toggles to the switch-blade assembly so that the opening and closing of the switch is not dependent on the speed of operation of the handle, this being effectively a spring loading device. The oil switch is a load-break device with fault-making capability.

A development of the oil switch is the switch-fuse, a similar equipment but with the addition of striker fuses in series with the switch blades, the complete assembly being housed in the oil tank. The fuses provide short-circuit protection, enabling the device to be used in place of the more expensive circuit-breaker (Fig. 9.13).

One of the main disadvantages of earlier generations of switch-fuses was that fuse renewal was a relativley long and complicated process, entailing isolating and earthing the equipment and then lowering the oil tank by means of a pulley or jacking arrangement before the fuses were accessible. With modern units, the mechanism is mounted at the top and the operation of the handle raises both the lid and the switch-fuse complete with fuses after full isolation has been effected. In all cases, access to live sections of the units cannot be gained until the necessary precautions have been taken.

Fig. 9.13 Yorkshire Switchgear Ltd Type FS-A0 200 A extensible outdoor automatic oil switch suitable for use at 15.5 kV and prospective short-circuit currents up to 21.9 kA.

The simplicity of design and operation and economy in cost of such units makes them suitable for the smaller high voltage installation.

Air circuit-breakers

The general description of the low voltage air circuit breaker described earlier in this chapter is equally applicable to the high voltage unit; the differences lie in the weight, size and clearance distances between the three phases and to earthed metal. These are, in fact, generally greater than in any other type of circuit-breaker and, as this type of breaker is usually far more expensive than others, it is less popular than other devices in commercial and industrial installations.

Table 9.1 summarizes the fields of application of the protective devices discussed. It should be noted that on some large commercial installations,

Table 9.1 Typical applications for fuses and switchgear.

	Domestic	Commercial	Industrial	Supply industry
Low voltage equipment				
Semi-enclosed fuse	Yes but out-moded	No	No	No
HRC fuse	Yes	Yes	Yes	Yes
MCB	Yes	Yes	Yes	Yes
MCCB	Yes (rare)	Yes	Yes	Yes
ACB	No	Yes	Yes	Yes
High voltage equipment				
Fuse	—	Yes (rare)	Yes	Yes
OCB	—	Yes	Yes	Yes
VCB	—	Yes but costly	Yes	Yes
SF_6 circuit breaker	—	Yes but costly	Yes	Yes
Min. oil circuit breaker	—	No Not popular in the UK	Yes	Yes
Ring main unit	—	Yes (rare)	Yes	Yes
Oil switch	—	Yes (rare)	Yes	Yes
Switch-fuse	—	Yes (rare)	Yes	Yes
ACB	—	Yes	Yes	Yes

e.g. multi-storey office blocks or shopping complexes, all types of high voltage fuses and switchgear are employed with the exception of the minimum oil circuit breaker.

For detailed technical information on protective devices, reference may be made to the *Handbook of Electrical Installation Practice* (second edition), E.A. Reeves (editor), published by Blackwells.

Short-circuit calculations

The two main functions of protective devices are to operate under both overload and short-circuit conditions, although these will also cover earth-fault currents of a similar level unless other more sensitive equipment is included in the installation, i.e. residual current devices.

The installation of HRC fuses, MCBs and MCCBs, provided they are suitably graded for the load currents, includes overload protection as an inherent feature at 1.45 times the rated current of the device, but short-circuit levels must be calculated from the total impedance of the system and examples are given below.

High voltage calculations

In high voltage networks the MVA method of determining the symmetrical short-circuit current is usually used and switchgear is rated in MVA.

$$\text{Fault current is therefore} = \frac{\text{MVA} \times 1000}{\sqrt{3} \times V_L}$$

Resistance of the windings in transformers, generators or reactors may be insignificant compared to their reactance. In these circumstances the fault power can be determined by the percentage reactance method where:

$$\% \text{ reactance } (X) = \frac{\text{voltage drop due to } I_b \text{ in } x \times 100}{V_L}$$

I_b being the full load current, V_L the line voltage and x the reactance of the windings.

Transformers and generators are likely to have different ratings which require each item being brought to a common base for calculation purposes. Fortunately, any value can be chosen as the base value.

Example 1

Figure 9.14 shows two generators, both of different ratings and percentage reactance. To enable calculations to be carried out a common base has to be chosen, this could be the size of the largest machine, but for this example 50 MVA will be chosen as the base MVA.

Working

Changing the percentage reactance of the 20 MVA machine:

$$\text{New } \%X = \frac{50\,\text{MVA}}{20\,\text{MVA}} \times 10\% = 25\%$$

Changing the percentage reactance of the 30 MVA machine:

$$\text{New } \%X = \frac{50\,\text{MVA}}{30\,\text{MVA}} \times 5\% = 8.3\%$$

To calculate the percentage reactance at the fault, the percentage reactances worked out to a common base can now be treated as resistance in series or parallel. In the above example the generators are in parallel therefore:

$$\% \text{ reactance at the fault} = \frac{25 \times 8.3}{25 + 8.3} = 6.23\%$$

$$\text{Fault power} = \frac{\text{Base MVA} \times 100}{\% \text{ reactance}} = \frac{50 \times 100}{6.23} = 802.6\,\text{MVA}$$

$$\text{Fault current} = \frac{\text{MVA} \times 1000}{\sqrt{3} \times V_L} = \frac{802.6 \times 1000}{1.732 \times 11\,\text{kV}} = 42\,127\,\text{A}$$

When the fault on the system is not symmetrical, the calculations become more involved and symmetrical components is then used to calculate the fault current.

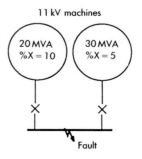

Fig. 9.14

Low voltage calculations

As far as the low voltage network is concerned the fault level on the high voltage side of the transformer can be ignored, since this would only slightly reduce the calculated fault level on the low voltage side of the system. The ohmic method of calculation for fault currents is used since the resistance of conductors will play a significant part in the calculations. On large installations the resistance and reactance of the transformer will be required and this is usually expressed in the form $R + jX$. The resistance and reactance of cables is expressed in the same way.

At the point that the fault current is required to be known, the resistances in series are added together, the reactances in series are added together and the impedance is found from the formula:

$$Z = \sqrt{(R_1 + R_2 + \ldots R_n)^2 + (X_1 + X_2 + \ldots X_n)^2}$$

Two calculations are required when carrying out short-circuit current calculations. The first to determine that the switchgear, starters equipment etc., will withstand the maximum peak fault current that could occur and the second to determine that the conductors or cables have sufficient withstand capacity. There is a subtle difference between the two calculations, this being the temperature at which the resistance of the conductors and windings is taken when carrying out the calculations.

When checking that switchgear or equipment will withstand the peak current likely to flow, the resistance of the conductors should be taken at the lowest temperature they could be operating at when a fault occurs. This will then give the maximum current the equipment etc., has to withstand. For example, the resistance of a transformer winding when cold (no load on it) would be taken at 15/20°C.

Where the supply is already available, and it is certain that it does not exceed 20 kA, the actual fault level can be determined by using the IMPSC instrument illustrated in Fig. 19.1. This value can then be used to check that equipment is of the correct rating for the fault current available.

When checking that the cables have sufficient withstand capacity, the resistance of the conductors must be taken at the average of the operating temperature of the conductor, and the limit temperature for the conductor's insulation. For example, ordinary PVC-insulated copper conductor cable has an operating temperature of 70°C with a limit temperature of 160°C for the insulation, the average is 115°C; this is the temperature used for the conductor resistance in calculations.

Most manufacturers' catalogues now give the resistance of conductors at 20°C for various types of conductor material. For the impedance of armoured cables, MICC cables, conduit and trunking as well as the resistances at the design temperature, specialized books dealing with the wiring regulations (listed at the end of this chapter) have to be consulted.

The resistance tables for copper and aluminium are now based on a simplified formula given in BS 6360, which uses a resistance-temperature coefficient of 0.004 per °C at 20°C. The resistance at any other temperature being calculated from the formula:

$$R_f = R_{20}(1 + \alpha (t_f - 20))$$

where R_f is the final resistance, R_{20} is the resistance at 20°C, α is the resistance temperature coefficient for the conductor material (0.004 for copper and aluminium) and t_f the final conductor temperature. Where the final temperature is 115°C then $R_f = 1.38\ R_{20}$.

Where the conductors are larger than 35 mm² reactance has to be taken into account, but no adjustment for temperature is made, since reactance is unaffected by temperature.

To enable the worked examples to be followed easily the following resistances of copper and aluminium are given at 20°C per 1000 m

Conductor size	Copper	Aluminium
2.5 mm²	7.41	—
35 mm²	0.524	0.868

Example 2

A final circuit to a 240 V × 1000 W floodlight consists of a 30 m long, two-core 2.5 mm² PVCSWA & PVC armoured cable. If the phase impedance up to the distribution board is 0.1 Ω what will the single-phase prospective fault current be:

(a) At the lighting point?
(b) At the distribution board?

Working

Impedance of 30 m of 2 × 2.5 mm² PVCSWA & PVC cable = 30 m × 2 conductors × 7.41 × 1.38 × 10⁻³ = 0.614 Ω.
This must now be added to the external phase impedance at the distribution board = 0.614 + (2 × 0.1) = 0.814 Ω.

Prospective short-circuit current at the lighting point $= \dfrac{240\,V}{0.814\,\Omega} = 294.8\,A$

Single-phase prospective short-circuit current at the distribution board $= \dfrac{240\,V}{2 \times 0.1} = 1200\,A$

As far as three phase circuits are concerned the minimum fault current will occur when the fault is between two phase conductors. The maximum fault current occurs with a symmetrical fault i.e., a fault across all three phases or across the three phases and neutral.

Example 3

A 415 V three-phase circuit comprising 100 m of 3 × 35 mm^2 PVCASA & PVC cable having aluminium conductors feeding a 45 kW motor is fed from a distribution board where Z_p at the bus bars is 0.1 Ω. Determine the minimum and maximum short circuit current at the motor.

Working

Impedance of one phase conductor to the motor:

(a) For minimum fault current $Z_p = 100 \, m \times 0.868 \times 1.38 \times 10^{-3} = 0.1198 \, Ω$

(b) For maximum fault current $Z_p = 100 \, m \times 0.868 \times 10^{-3} = 0.0868 \, Ω$

Minimum prospective short-circuit current I_{pp} will occur between two phases.

$$\text{Minimum } I_{pp} \text{ at the motor} = \frac{V_L}{2 \times Z_p} = \frac{415 \, V}{2 \times (0.1 + 0.1198)} = 944 \, A$$

$$\begin{array}{l} \text{Maximum short-circuit} \\ \text{current at the motor } I_p \end{array} = \frac{V_p}{Z_p} = \frac{240 \, V}{(0.1 + 0.0868)} = 1285 \, A$$

Calculations concerning earth faults are carried out in a similar manner except the current is only limited by the phase earth loop impedance. The information required includes the earth loop impedance external to the circuit being considered, designated Z_E, as well as the impedance of the circuit's phase conductor and protective conductor.

Since a minimum fault current must flow for the phase conductors to be protected, it follows that if the earth fault current is less than the short-circuit current, the suitability of the phase conductor must again be checked.

Protective discrimination

In commercial and industrial installations, a number of protective devices are installed in series, e.g. final circuit distribution boards, submain switchboards and main switchboards, and these are set to ensure that, if a fault occurs, the protective device nearest to the fault operates first to avoid unnecessary shut-downs; this is referred to as discrimination. Additionally, it is essential to ensure that a protective device operates before a fault current reaches a level that will raise the temperature of the associated cable insulation to the extent that its effectiveness is destroyed. This will occur long before the actual conductors fail. The temperature limits for standard cables are given in the IEE Tables 4D1 etc.

To obtain effective discrimination it is necessary to refer to manufacturers' data sheets and, preferably, avoid installing devices from different makers as, although characteristics may be similar, they can differ sufficiently to prevent discrimination.

The short-circuit rating of a protective device is measured in terms of 'energy let-through' or I^2t, that is, the amount of energy that can be safely carried by the protective device under short-time fault conditions, and it is this quantity that must be compared to ensure adequacy. Figure 9.15 indicates typical I^2t characteristics for a selection of HRC fuses; in order to obtain discrimination between fuses in series, the total operating I^2t of the smaller fuse down-line must be less than the pre-arcing I^2t of the preceding one in the circuit. In the examples shown, the 100 A fuse does not discriminate against the 125 A and, therefore, the lower rating of 80 A is the maximum that may be installed down-line. A similar situation exists with 40, 35 and 32 A fuses and with the 35, 32 and 25 A.

From the above examples it will be appreciated that, in many cases, the rule-of-thumb ratio of fuse ratings of 2:1 provides a high safety factor for HRC fuses and is unnecessarily restrictive. It also leads to unnecessarily

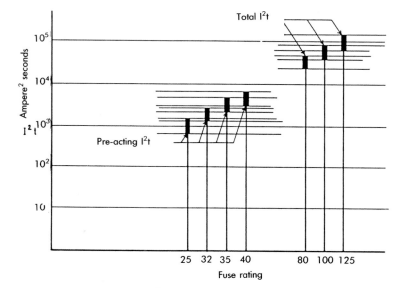

Fig. 9.15 I^2t characteristics of HRC fuses.

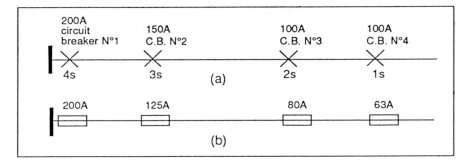

Fig. 9.16 Discrimination diagrams by (a) time delay or (b) current rating.

large cables being used. Discrimination may also be achieved by time-delay as opposed to protective device rating as shown in Fig. 9.16.

Although, logically, it might be assumed that a lower rated protective device will operate in advance of a higher rated one in series in the same circuit, this may not always be the case with preset units such as MCBs. Referring to Fig. 9.17, which is a reproduction of part of Fig. 5 of the IEE Wiring Regulations − Time/current characteristics for Type 2 miniature circuit-breakers to BS 3871 − it is evident that there is a finite minimum tripping time, of approximately 10 ms, for every rating. From the horizontal axis it will be seen that, with a fault current in excess of 500 A, every rating of this type of MCB will trip at virtually the same time. Consequently, the possibility of sudden high fault currents must be taken into consideration when it is proposed to install these devices in series.

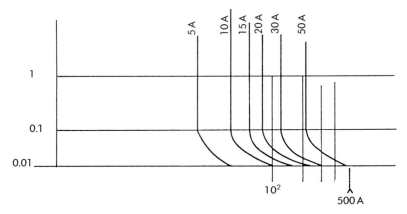

Fig. 9.17 Time/current characteristic curves for miniature circuit-breakers − based on Fig. 5 of the IEE Wiring Regulations.

With relay-operated protective devices this difficulty does not arise, as time-lag is also introduced by means of oil dashpots or induction relays, however, this increases the disconnection time as far as protection against indirect contact is concerned. It is usual, however, in this case to employ inverse-time characteristics to ensure rapid disconnection of a faulty circuit in the event of an excessively high fault current.

One aspect of discrimination which must be taken into consideration on ring main systems is that, depending upon the point at which a ring is open, current may flow in either direction through the cables. It is necessary, therefore, to apply discrimination both ways at each protective device.

References for impedance of conduit, trunking and armoured cables

Jenkins, B.D. *Electrical Installation Calculations*, Blackwell Scientific, Oxford, 1991.
Marks, T.E. *Handbook on the IEE Wiring Regulations* (3rd edn), William Ernest, Nottingham, 1992.

Chapter 10
Cable Support Systems

For the smaller installations, conduit and trunking will often adequately cover all the requirements and, for final circuits, they are quite suitable in many of the larger ones also. As cables increase in size, however, they become less flexible and more difficult to install in such confined spaces and, therefore, other means (described below) are used to support them which give the additional benefits of greater accessibility and reduction of grouping problems.

Cleats and brackets

For a single cable, or a relatively small number of cables, cleats and/or brackets are used which are fixed to an existing structure such as the metal framework or walls of a building or, where suitable support is not available, to channels or similar steelwork installed for the purpose. This latter is essential for crossing large areas of corrugated steel or fragile sheeting as in industrial premises or, in some cases, the roofs of buildings such as supermarkets.

There are a large number of commercially produced types of cleat available which are suitable for both a single cable or a small number of cables, and a triform type for clamping three single-core cables in close proximity to each other. For very large cables it is advisable to cleat each one separately, rather than use multiway cleats, as they are then more easily handled.

Long horizontal runs of cable are adequately supported and more easily installed on J-type brackets as it is then only necessary to run out the cable and lift it on to the brackets, whereas with clamps the bases have to be fixed in place and the clamps completed after the cable has been seated on the base.

Whether cleats or brackets are used, care should be taken to ensure that these are not too widely spaced as this will lead to cable sag and, eventually, damage to the cable sheath and possibly to the conductors.

As cables supported in this manner receive no mechanical protection from the support system, it is more suitable for armoured cables or in situations where there is only a remote possibility of damage from other sources such as forklift trucks and similar. Direct fixing to the building structure is also not recommended if it is subject to heavy vibration.

Further information on cable support is given in Appendix I of the IEE Guide to Selection and Erection in which Table I2 deals with maximum spacing of cables up to 40 mm overall diameter. For larger cables, the appropriate information is always to hand from the manufacturers.

Cable trays

When relatively small numbers of less heavy cables are required there are many types of cable tray available, produced from perforated steel or plastics sheeting and in various forms and strengths. Plastics (heavy-grade PVC or similar) tray is common for light duty or in corrosive conditions although plastics-coated metal tray is also produced for such situations; if a coated tray is used, all cut ends must be treated with rubberized paint or similar, otherwise corrosive substances will attack the bare steel.

The lighter types of tray are provided with upturned sides while, for additional strength, rolled-edge tray is usually employed. In all cases, accessories may be used to provide changes of direction although small longitudinal sets are often accomplished by cutting the edges and bending the tray; if this method is used, reinforcement may be required to retain the strength of the tray (Fig. 10.1).

Ladder-racking

For the heavier duties such as in steelworks, coal-mining, etc. or any situation where extremely large cables are to be installed, particularly on structures subject to heavy vibration, ladder-racking is more suitable than tray for multiple cable runs as it is manufactured in much greater strengths and is generally far more substantial than tray (Fig. 10.2).

Ladder-racking, in appearance, resembles the well-known steel or aluminium builders' ladder (from which the name was derived) and is formed from the same types of tubular material. Unlike tray, which provides continuous support for cables, the 'rungs' are spaced (approximately 300 mm, depending upon the maker) but they are sufficiently close to make further support unnecessary. Again, various accessories are available to facilitate changes in direction and, as with tray, cable clamps are not usually necessary except on vertical runs or where the racking is installed on edge to negotiate, for example, a narrow opening.

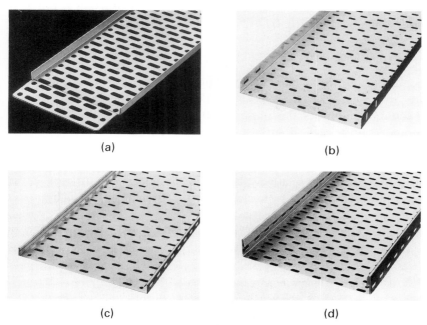

Fig. 10.1 Types of tray (a) Light duty cable tray (Admiralty pattern); (b) heavy duty cable tray; (c) medium duty return flange cable tray; (d) heavy duty return flange cable tray.

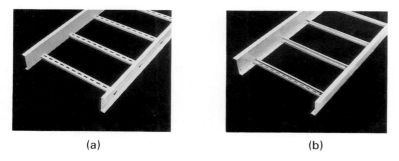

Fig. 10.2 Types of ladder rack; (a) medium duty; (b) heavy duty.

As ladder-racking (and tray) is supplied in various standard lengths, there is little difficulty in assembling any required overall length from the manufactured standards or, probably less costly, by cutting to length on site.

Possibly one of the greatest advantages of ladder-racking is that, due to its strength, it is possible to span far longer distances than with other forms of support system and, therefore, the amount of fixing required is considerably less (Fig. 10.3).

Special considerations

When cables are installed in dusty atmospheres, especially in hazardous areas, care should be taken to ensure that the minimum amount of dust will settle on them; for this reason cleated cables should be spaced apart if mounted on a vertical surface, and those on tray and ladder-racking segregated. Spacing also assists cooling and, therefore, can avoid cable derating. Also, whenever possible, the support systems should be so placed that they are relatively accessible for cleaning purposes, more necessary with tray than with racking as the former has a much greater surface area.

A further important aspect, if the tray or ladder-racking forms part or

Fig. 10.3 BICC Pyrotenax mineral-insulated cables installed on Vantrunk cable tray at Glaxo (UK) Ltd, Annan.

all of the protective conductor system, is that care must be taken with the jointing of the various sections to ensure electrical continuity as these are often butt-jointed with bolted fishplates.

Builders' work ducts

Although, as already indicated, there are different methods available enabling cables to be installed at high level, many situations arise which make it preferable for them to be concealed yet still be accessible. This is normally the case in buildings such as multistorey office blocks, super-markets, department stores, and even high-rise dwellings, where appearance is important. In industrial complexes there are many other services above ground which have to be concealed and for which constant access is needed or, as in the food manufacturing industry, where it is necessary to keep dust-harbouring surfaces to a minimum. Further, large cable installations add considerable weight on a structure and, even with modern equipment, are not easy or economical to install at high level.

In the above instances and, no doubt, many others, it is preferable to use builders' work ducts for this purpose, these taking several forms as described below.

Single-way ducts

If only a small number of cables are to be installed on a particular run, a single-way, high-glazed finish earthenware duct or fibre duct manufactured specifically for cable-ways is installed by the builder during the construction of the foundations and floor slab or along the required route between buildings etc. Earthenware ducts are of the order of 1 m in length with spigot and socket ends, while fibre ducts are usually some 3 m long and are joined together by a double-ended socket and neoprene sealing rings. Both types are available in large diameters (compared to conduit), the smallest being approximately 75 mm. As a variation on the earthenware type, split ducts are obtainable which permit the installation of cables on the bottom half before the top is seated and back-filling commenced. This type simplifies the installation of the very large, less flexible cables.

It is more usual to install the single-way duct for high voltage or the main low voltage cables, i.e. those that are the least likely to require replacement under normal circumstances although, because of their capacity, it is possible to install several cables in the one duct; this may, however, present difficulties in the event that a cable has to be added or an existing one withdrawn. To provide additional facilities it is preferable to install a spare duct or, in the case of earthenware, a manufactured nest of four or six ducts.

From the descriptions given, it is clear that the fibre duct provides a more watertight run than the earthenware (although the latter may be sealed with bitumen or similar) but, with modern PVC, EPR (ethylene propylene rubber) or HOFR (heat, oil and flame resistant) sheathed cables, sealing of the ducts along the route may not be necessary but should be applied at the ends.

As with all types of cable enclosure, fittings are available to allow route diversions and upstands of sufficient radii to suit the cable requirements.

As it is the normal practice for the builder, who is not always well versed in cable installations, to install ducts, the designer/installer must, before work commences, provide full information of his requirements regarding route, depth, any additional protection for the ducts and exact terminating points. With the closed duct also, arrangements should be made with the builder to insert a draw-wire as the duct-laying proceeds and to clean out any sealant or other foreign matter to avoid the expense of employing specialized equipment for this purpose after installation.

On large industrial installations it is often required to install long runs of cable in ducts between buildings or across a factory floor and, although mechanical capstans etc. are capable of drawing these through, it is advisable to provide intermediate pits or chambers along the route to avoid excessive strain on, and damage to, the cables. If multiway ducts are installed for a large number of cables, pits should also be provided at diversions.

Cable trenches

As an alternative to the use of single-way ducts, cables may be installed in trenches which on new buildings, as with ducts, are built by the main contractor. In open ground these are usually simple excavations of the required depth. Before cables are installed, it is recommended that a 75 mm layer of sifted sand is placed on the bottom of the trench on which the cables are bedded; a further 75 mm of sand is then added, and interlocking clay tiles laid over. Opinions differ regarding the necessity for tiles but, with major low voltage and all high voltage cables, it is strongly advised to use them because, although an area may be considered safe initially, circumstances may arise in the future which alter the situation. The trench is then partially back-filled, a coloured plastics tape laid along the length of the cable as a warning, and filling and compaction of the trench completed.

Across floors, trenches are formed in brick or concrete with recessed edges to allow a load-bearing cover to be installed over the trench. It is usual to fix brackets along one or both sides of such trenches on which cables are laid and cleated or, for small cables, tray installed. The cable layout in a trench should be carefully planned so that the first cables to peel off from the trench do not reduce accessibility to the remainder. The overall size of a trench is governed by the number of cables to be installed, with spare capacity and sufficient width to provide access to the lower runs; for the size of commercial or industrial installation considered, a 600 mm by 600 mm trench would normally be adequate.

To avoid sharp bends in cables at branch trenches, it is usual to chamfer the corners of the junction.

Vertical ducts

As previously mentioned, if concealment of cables, tray, trunking, rising mains, etc. is preferred or necessary, vertical ducts are included in a building, either in the form of a chimney (which may then be classified as a fire compartment if of adequate fire rating) or, when frequent branches, switchgear panels or metering assemblies are required as in multistorey flats and office blocks, in the form of a series of cupboards one above the other on each floor.

In large multistorey open-plan offices, a central position to avoid long wiring runs may be required which cannot be accommodated in the actual building structure. In such cases it may be feasible to install false columns in which to locate electrical services.

It should be noted that, in all cases where ducts are not classified as fire compartments, the appropriate authorities demand that fire barriers be installed between the different levels.

Walkway service ducts

In many commercial and industrial buildings numerous different services are required, such as hot and cold water, steam, 'fridge' lines, compressed air, ventilation ductwork and process pipework, in addition to the electrical services and, very often, a large amount of otherwise unused roof space is available which is adequate for all the required services. However, while each service may be installed separately, this will provide an unsatisfactory and inaccessible mass of pipework etc. For such cases, a walkway service duct provides a satisfactory solution. On this type of walkway, all the services are accommodated on the sides supported on brackets, tray or ladder-racking, as described above, and leaving adequate space for service personnel to work as required in safety. These walkways are extremely substantial and may be supported from the main structure of a building, themselves being used at times for the intermediate suspension points of false ceilings in, for example, single-storey supermarkets or, in open roof areas, for the suspension of luminaires. Similar walkways are also often provided below ground-floor level, particularly on large projects such as power stations, and, in these, cables, tray, etc. are sometimes suspended from the ceiling (Figs 10.4 and 10.5).

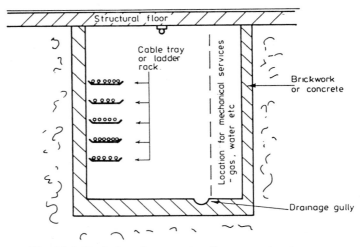

Fig. 10.4 Typical underground walkway service duct.

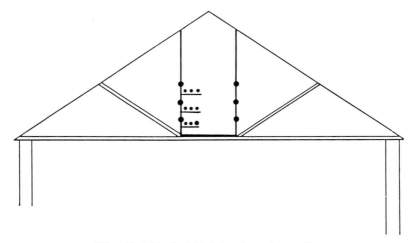

Fig. 10.5 Typical high-level service walkway.

Duct sealing methods

Every type of aperture through the structure of a building provided for the installation of cables also provides a route for vermin, water or fire. In most cases, sealing of the apertures is advisable and, in some, compulsory in order to comply with statutory regulations or local authority by-laws.

Single and multiway ducts provide the least problems for sealing as they are usually built into the structure of a building and concreted, leaving only the ends to be dealt with, a process which may be effected by packing round the cables with fibrous material and finishing with compound or malleable plastics. As an alternative, where a duct is used for a single cable or a small number of cables, a plug of timber or similar is prepared and drilled for the cables to pass through. Sealing is then carried out as described. In addition to these methods, it is possible to obtain manufactured, flexible caps which are simply threaded over the cables and the duct end. Builders' trenches, tray and ladder rack through a structure, carrying a number of cables, obviously present rather more problems than ducts, particularly in hazardous areas. Apart from the practical difficulties, there is always the possibility that some alteration will be required at a later date for which provision must be made. Fortunately, the modern smooth sheathing materials now used for cables allow seals to be more effective.

One of the methods adopted, particularly for flameproof installations, is to terminate the support system on each side of the aperture, plate and drill for each cable and, as they are installed, thread a sealing gland over each and tighten the glands when the work is finished.

A current, and possibly more effective, method has been developed in the USA and Sweden. This entails temporarily blocking the aperture after the installation has been completed and applying, by means of pressure nozzles, chemical compounds which react to form a pliable mass around the cables and supports. The finished seal has the advantages that it is relatively easy to penetrate if further cables are required, does not have to be broken out completely for alterations and, of the greatest benefit, provides an acceptable fireproof barrier even when subjected to intense heat.

This method has been approved and used by many major authorities throughout the world, after exhaustive tests. It is emphasized, however, that a number of such methods of sealing are available using different products and characteristics; it is essential, therefore, to fully discuss the requirements with the specialist manufacturers and/or installers before proceeding.

With regard to the permanent plating of apertures, the most commonly used material was asbestos sheeting which was easily worked and had fireproofing characteristics. Due to the possible health hazard, this is now unpopular and prohibited entirely by health authorities, although the evidence appears to indicate that only blue asbestos is harmful. There are, however, many suitable materials available for plating including plastics, timber and steel, while many types of fabricated panel used in modern building construction can also be used for this purpose.

A method of sealing introduced by Hawke Cable Glands Ltd, claimed to be the first UK designed and manufactured system, consists of a frame which is secured to the structure over an aperture through which cables are threaded. Shaped blocks of flame-retardant intumescent elastomer are packed around the cables within the frame and the total assembly of cables and packings is then compressed by a clamping device attached to the top of the frame.

The packings are sufficiently elastic to overcome irregularities in the cable sheaths and, should the latter be destroyed by excessive heat or fire, the elastomer expands to fill any voids created and maintains the integrity of the fire barrier.

The frame is sufficiently large to allow a multiplicity of cables of differing sizes, and it is relatively simple to install additional cables at a later date by unclamping the assembly, removing make-up packings and threading the extra cable through the remaining space. Further packings are then inserted around the added cable and the whole assembly reclamped.

As with any form of sealing barrier, care must be taken to ensure that gaps between the frame and structure are effectively sealed as, otherwise, the purpose of the barrier will be defeated.

With the advent of the 16th Edition of the IEE Wiring Regulations the requirements for fire barriers are now more rigorous, which will make the supply of suitable materials more inviting to manufacturers.

Chapter 11
Enclosed Wiring Systems

In the earlier days of electricity supply, direct current (d.c.) at low voltage was commonly employed, generators often being driven by prime movers (water wheels, steam engines, etc.) previously used for shaft and pulley assemblies. Consequently, many of the original installations consisted of single-core cables supported in cleats. With increasing awareness of the possibility of danger, the necessity for greater protection created the demand for enclosures such as conduit and, later, trunking of which there are many different types now available to suit any situation, as described below.

Cable trunking

For general use, cable trunking is now available in various materials such as steel, PVC, aluminium and phenylene oxide (Noryl), in a wide range of sizes of both square and rectangular cross-section. The trunking consists of a three-sided fixed section with internal or external flanges on the open side to permit the fixing of the lid after the wiring has been installed; when necessitated by the environment, trunking with gasketed lids may be obtained. Support brackets of different types are available from all trunking manufacturers who, because of the varying strengths of construction materials, are always willing to provide the maximum loading (weight) that a given trunking will carry without noticeable sag. Appendix I of the IEE Guide to Selection and Erection also provides information on the maximum spacings permitted for cables, conductors and wiring systems. Often when fixed to a flat surface, plastics trunking is simply drilled and installed in place by screws and wallplugs.

Steel cable trunking is supplied in various standard lengths with provision for slotting together and bolting to maintain electrical continuity for bonding. If required, trunking is available with pin supports at regular intervals for separating circuits and, where it is essential to completely segregate wiring, such as safety services and extra-low voltage, continuous barriers are provided.

Various finishes, such as paint-dipped, galvanized or sherardized are possible and the installer should ensure that, if arduous environmental conditions exist, these are specified.

Plastics trunking is usually jointed together with proprietary connectors and adhesive, although mechanical systems are available. If adhesive is used, it should only be applied to one end of the connector, the other being coated with a nonsetting mastic or grease as this allows for expansion or contraction. As plastics are not conductive, a protective conductor is essential.

A range of accessories is available to retain cables in position or to segregate different circuits, for internal and external bends, for tee connections, etc.

In the UK the manufacturing standard is BS 4678: Part 1 for steel trunking and Part 4 for trunking made from insulating material.

An individual type of trunking intended for specifically hazardous conditions in industry and commerce is produced by Van Trunk Engineering Ltd and is constructed around their ladder support system. The ladder-racking is enveloped in a high grade ceramic fibre or mineral wool which is then protected by sheet steel. The trunking contains sachets of treated water laid on the thermal insulation and attached to the walls which are designed to allow water vapour to escape at a temperature of approximately 110°C. These sachets absorb excessive heat, thus reducing the temperature rise within the trunking. It is claimed by the manufacturers that the system (Firetrunk) will withstand temperatures of 1100°C for up to three hours. The relatively high cost of Firetrunk precludes its use for normal purposes but many small industrial processes may include areas which are particularly susceptible to fire risk and, therefore, require such additional protection.

Floor distribution systems

For low level distribution, particularly for open-plan offices and similar, two types of floor system can be used, both taking a shallow rectangular form with or without continuous barriers. Generally, however, the former is preferable to provide segregation facilities for the different services, e.g. power, data, communications.

To provide easy access for alteration to existing cables or for additional circuits, the former flush-fit ducting is the better choice. This type is provided with flanges which permit floor screeds to be finished up to it without unsightly gaps and allowing for the final finishes without projections. Ducting lids, which are continuous, are formed in heavy-duty metal and may be recessed to take a floor finish such as carpet or plastic tiles where appearance is important.

The alternative is to use hollow-section ducting built directly into the floor structure and since it is in box form, there will be no access to the ducting except at floor outlet and junction boxes. Separate ducts can be installed for segregation purposes, cross-overs at junction boxes cause difficulties and are therefore more suited to single runs.

Accessories are available which enable outlets to be fixed onto the junction boxes in the trunking built into the floor, or for pedestal-mounted pattern or flushfloor fittings complete with hinged lid for access to the outlets to be mounted onto the flush fit trunking.

Careful planning is required for the layouts of both types of distribution system to ensure that all foreseeable requirements will be met, particularly with regard to underfloor ducting as, although most types are provided at regular intervals with screwed turrets and plugs to facilitate the installation of further outlets at a later date, these may not be where required and drilling into the ducting after wiring is not to be recommended. However, if the ducting is installed in a grid formation, this situation is unlikely to arise. See BS 4678: Part 2 for the manufacturing standards.

With both forms of ducting, it is essential to co-ordinate with the architect or builder regarding the floor construction as, otherwise, it may be found that reinforcement or the depth of floor screed prevent satisfactory installation.

To avoid some of the problems associated with the above types of system, Gilflex provide a carpet trunking for fixing to a finished floor, which has a total depth of 9.6 mm (Fig. 11.1). It is complete with a snap-on overlapping lid which, when in place, forms a retainer for abutting carpet.

Figure 11.2 shows the three types of enclosure.

Skirting trunking

In many cases, skirting trunking provides a suitable alternative to built-in wiring systems and is obtainable in steel or plastics with various external contours and cross-sections. Many accessories are marketed which allow for internal and external bends, for outlets and for junctions. The trunking is again two-part construction, so that the base with cableways is installed first and the lid added after wiring has been completed. For economy, skirting trunking is intended to take the place of the usual skirting boards

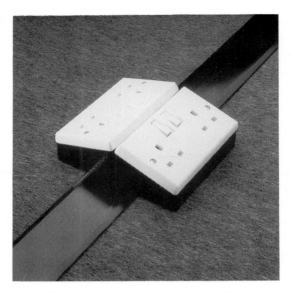

Fig. 11.1 Two-gang back-to-back arrangement of pedestal units on Gilflex carpet trunking.

and is available in multicompartment designs to provide segregation facilities.

The problem of crossing doorways etc. may be overcome by the use of an underfloor fitting or, less economically from the wiring aspect, by traversing around the complete frame. Some designs of trunking are suitable for mounting at dado or skirting height, thus eliminating the problem of matching-up a skirting and dado system (Figure 11.3).

Mini-trunking

For domestic or similar small installations, manufacturers have developed plastics mini-trunking systems which are similar in form to cable trunking but of less obtrusive cross-section, ranging from 16 mm to 75 mm wide by 12 mm to 30 mm deep. In the smallest size, Gilflex Homeline is a good example (Fig. 11.4). There are numerous accessories for bends, junctions and outlets and, with the exception of the outlets which are usually surface mounted, a complete installation can be installed quite inconspicuously by close fitting to skirtings, picture rails and door architraves. Because of the small section, runs on walls or across ceilings can be used without spoiling the aesthetics of an area.

A further advantage of mini-trunking is that it can be used in conjunction

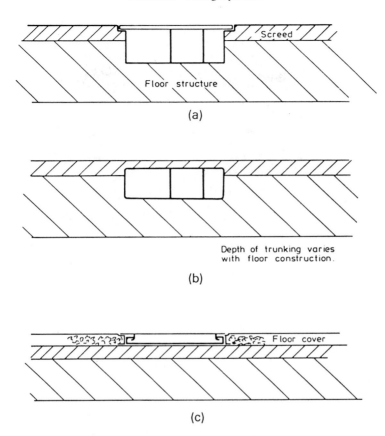

Fig. 11.2 Diagrams of floor ducting installation methods: (a) flushfit ducting, (b) built-in ducting, and (c) carpet-depth mini-trunking.

with cove-trunking for running cables to high-level fittings such as wall lights; the cove-trunking provides a cableway at high level with mini-trunking drops to the fittings.

Installation of mini-trunking can be effected by means of suitable adhesives on flat surfaces or, preferably, with wallplugs and woodscrews. Wiring the trunking presents no problems as continuous lids are provided which simply clip on to the base.

Mini-trunking and cove-trunking are particularly suitable for areas which may be subject to changes of layouts, or for rewiring, to avoid major unheavals in addition to new installations. The simplicity of installation and the degree of accessibility provided by these systems can reduce labour costs tremendously.

Fig. 11.3 Skirting trunking from Gilflex available in two- or three-compartment designs for telecommunications and power cabling providing full segregation. The three-compartment design for skirting or dado applications is also available in white.

Fig. 11.4 Gilflex Homeline mini-trunking in a typical boarding house bedroom at Harrow School. Neat joints between trunking and terminal equipment and unobtrusiveness are features of the installation.

Platform floor systems

The introduction of solid floors and the later development of the 'automated office' with its greater requirement for circuitry of different types encouraged the use of raised floors to provide a void suitable for housing power and communication services.

Platform floors provide an alternative floor distribution system to those described earlier, and are particularly suitable for computer rooms and similar places where data and other services may need to be changed on a fairly regular basis. Basically, the system consists of floor boxes which fit into movable floor panels with cable laying in the substructure floor. This cabling is best enclosed in conduit or trunking to ease wiring. Final cabling to an outlet box is usually long enough to allow the panel to be moved to alter the position of the outlets to meet new requirements. There are some designs of floor in which the floor panels are fixed, thus restricting the versatility of the system (Figs 11.5 and 11.6).

Service poles

Various types of service pole are available for use in open-plan offices which provide cable-ways for data and power outlets to isolated sites and reception desks, for column-mounted luminaires or to transfer cabling between floor and ceiling systems.

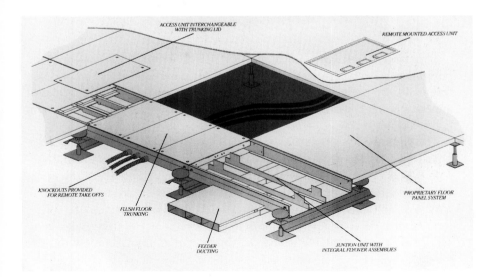

Fig. 11.5 Britmac type BRF flush floor trunking for raised floors.

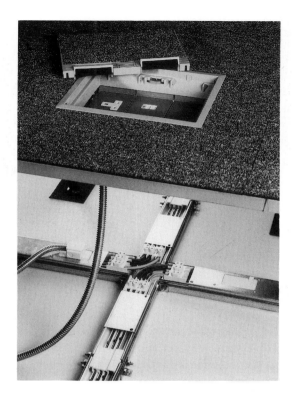

Fig. 11.6 Underfloor ducting installation. (Courtesy of MK Electric Ltd.)

The service poles are adjustable for height within limits and may be permanently fixed at the ends or simply retained in position by compression feet built into the poles, allowing for resiting if necessary. For the sake of appearance, they are designed to give the impression of being a structural feature (Fig. 11.7).

The capacity for low level socket-outlets and through-wiring of service poles removes the necessity for a large number of false columns in open-plan offices or similar areas, which would to some extent defeat the effect of spaciousness, although they are of much lower cross-section than structural columns. Consequently, to avoid the unsafe practice of a multitude of flexible cables across floors to areas remote from the poles, stub service posts are available to match, which extend approximately to desk height.

Obviously, a carefully designed combination of service poles and posts is capable of providing supply to many arrangements of service positions

Fig. 11.7 The Gilflex power-pole.

in offices and other commercial premises without being obtrusive or causing unnecessary obstructions (Fig. 11.8).

Conduit systems

Steel

The most popular form of enclosure for final circuit wiring in commercial and industrial installations has been steel conduit, although this has now been overtaken in many applications by plastics. Apart from the quality of materials and, since the late 1960s, metric sizing, very few changes have taken place. Steel conduit is manufactured with an enamel, sherardized or galvanized finish and in light and heavy gauge. For arduous conditions screwed conduit, as against the push-fit type, is usually preferred in the UK, although this choice is not universal. The largest diameter of conduit covered by the British Standard 6099 (which closely relates to international standards) is 32 mm (although 1.5 in conduct is available) which, although suitable for a large number of the smaller single-core cables, subject to other considerations, becomes less useful for larger

Fig. 11.8 Power Post. (Courtesy of MK Electric Ltd.)

ones, not only for the capacity but also because of the greater difficulty of drawing them into the system.

Heavy gauge conduit, whether screwed or welded, has for some years been accepted for use in hazardous areas, although special fittings are required for these. For detailed requirements of equipment suitable for such areas, reference should be made to BS 5345 and BS 5501. In domestic installations, oval conduit is used for droppers concealed under plaster but channel-type capping, which is a lighter gauge and cheaper, is more popular and is invariably used instead of conduit.

Plastics

When first introduced, plastics conduit had many disadvantages compared to steel: the material was mechanically weak, greatly affected by changes in temperature, did not retain sets, maintained combustion (and emitted toxic fumes) and tended to separate at joints. These problems have now been overcome and, in some respects, plastics conduit has many advantages over steel. It is much lighter and, therefore, easier to handle and install, provides a smoother surface for the drawing in of the cables, is not subject to corrosion and rusting, and the super high impact materials now used make it suitable for most applications.

Plastics conduit is produced in both light and heavy gauges, round and oval cross-section and metric diameters up to 50 mm with a range of accessories to suit all requirements.

One disadvantage that remains with screwed plastics conduit is that when lengths have to be cut, removing the preformed threads, the formation of a new screwed end depends upon the operative on site. This is easily effected with steel conduit but, with plastics, may lead to unnecessary weakening at the threads by cutting too deeply into the material. The most recent introduction is that of screwless fittings which avoid this problem and speed up installation. Some manufacturers produce fittings with a projection inside the socket which clamps the conduit when inserted while others prefer a push-fit with the application of a special adhesive to the exterior of the conduit end.

The applicable standard is BS 6099, Section 2.2 (earlier BS 4607).

Flexible

Several different types of flexible conduit are available, ranging from convoluted plastics to reinforced corrugated steel covered both internally and externally with self-extinguishing plastics, the last being the most appropriate for general use. It is particularly useful for final connections to machinery subject to vibration in place of the alternative methods of flexible cable or coiled mineral insulated copper cables (MICCs). One early disadvantage was that the conduit tended to break away from the terminations, but this has been rectified by the modern conduit glands to such an extent that, with certain restrictions, flexible conduit may even be used in hazardous areas.

As with rigid conduit, flexible conduit is available in several metric sizes and, additionally, can be supplied to required lengths with conduit glands already permanently attached. BS 731 deals with certain types of flexible conduit.

Handling of plastics

The term 'plastics' covers two broad groups of materials, referred to as thermosetting and thermoplastics, which have completely different characteristics from each other. For the installation engineer, unplasticized PVC, a thermoplastics, is the most important as it is the material used for the manufacture of the plastics conduit referred to previously, having the required properties for the application of relatively high strength, and good chemical and abrasion resistance; although fillers are more commonly used for thermosetting materials, natural fibres and other additives may be added to PVC in the initial mix to increase the strength.

For practical purposes, it should be noted that the material is affected by both extremely high and low temperatures, the first causing it to soften and, eventually, turn into a viscous liquid, while low temperature makes it brittle. This latter is particularly important when handling PVC cables and users are warned against installing such cables during very cold weather, although some designs are suitable for use down to −25°C and most can be used at −5°C. Plastics conduit is more often than not, however, installed within buildings where the conditions in the UK are unlikely to reach the extremes and, therefore, caution may only be required during transportation rather than installation. The effect of heat on PVC can be used with good results, if carefully applied, to prepare sets and bends in plastics conduit and so reduce the number of manufactured fittings and associated joints.

An excellent and detailed article on the properties of many plastics is included in *Newnes' Electrical Pocket Book* (21st edn), E.A. Reeves (editor), Butterworth-Heinemann, Oxford, 1992.

BS 6099 covers conduits generally, having partially superseded BS 4607. However, other standards such as BS 31 and BS 4568 may also apply, BS 4568 being the metric version of BS 31.

Chapter 12
Final Circuit Design

Although the IEE Wiring Regulations were first issued in 1882, the 15th Edition is the first to acknowledge the importance of design in detail by including this aspect of installation work in Part 1 — Scope, Regulation 11—1; this regulation also appears in the 16th Edition of the IEE Wiring Regulations as Regulation 110—01—01. Throughout this edition, much information was included to assist the designer, while not removing the responsibility for effective application or the requirement for sound engineering ability. Before final circuit design is commenced, it is essential to obtain as much information as possible about the whole project to include such details as usage, building dimensions, type of construction, required locations for equipment (i.e. workstations), escape routes, electrical loads and types of starter for plant not supplied by the designer/installer, and the presence of hazardous materials.

It is also necessary, as indicated in Chapter 1, to ascertain from the regional electricity company the type of supply to be provided and the prospective fault level and external impedance (Z_E) at the origin of the installation, and to agree the location of the origin of supply.

Design procedure

For every type of circuit, lighting, socket-outlets or power, the basic design procedure differs only in detail as, for example, between fixed equipment and socket-outlet circuits. The following is based on the requirements of the 16th Edition of the IEE Wiring Regulations:

(1) Calculate or otherwise obtain the design current (I_b).
(2) Based on I_b and the fault level, the type and nominal rating of the protective device is selected. Both fault level rating and I_n must equal or exceed the required figures.
(3) Calculate the current rating of the cable I_z by dividing I_n or I_b by the applicable correction factors (see Chapter 8).
  (a) Ambient temperature if other than 30°C.
  (b) Grouping factor from Table 4B1 or 4B2 of the IEE Regulations (but see Notes to Table).

(c) Thermal insulation factor (Reg. 523−04−01).

(d) Factor of 0.725 only if semi-enclosed fuses are to be installed for overload protection.

(4) From Appendix 4 select the appropriate cable rating for the type of cable to be installed. The rating must equal or exceed the calculated figure.

(5) Calculate the voltage drop by multiplying the appropriate tabulated millivolt drop value by the circuit length and I_b.

(6) Check the thermal capability of the circuit conductors at the minimum and maximum prospective short-circuit current (as explained in Chapter 9) by using the formula $I^2t \le k^2S^2$ where:

I = prospective short-circuit current.

t = permitted disconnection time.

k = constant from Section 434 of the IEE Wiring Regulations.

S = cross-sectional area of the circuit conductors.

The disconnection time t calculated from the above formula is then checked against the actual disconnection time of the protective device for the current I; the latter should be less than the calculated value.

(7) Depending upon the appropriate period of disconnection, 5 s for fixed equipment and 0.4 s for socket outlet or hand-held equipment circuits, select the maximum permitted earth loop impedance Z_S from the appropriate table for the type of protective device being used and deduct Z_E. The part of the system covered by Z_E is illustrated in Fig. 12.1 (a and b).

Deducting Z_E from Z_S leaves the amount of earth loop impedance left for the circuit Z_{inst}. Now work out the actual earth loop impedance of the circuit, if it is more than the allowed Z_{inst} a larger cable has to be installed, whose phase earth loop impedance does not exceed the allowed value of Z_{inst}. The resistance of the conductors is taken as the average of the conductor's normal operating temperature and the limit temperature for the conductor's insulation under fault conditions.

The disconnection time for socket outlet circuits can be increased to 5 s providing the impedance of the protective conductor does not exceed the value given in Table 41C for the type of protective device being installed.

Where the circuit feeds equipment is installed outdoors i.e., outside the equipotential zone, the disconnection time allowed for circuits feeding fixed equipment is 0.4 s and circuits for socket-outlets or hand-held equipment must be from a 30 mA residual current device (RCD).

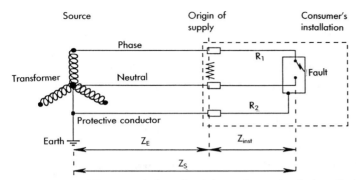

Fig. 12.1a Phase earth loop impedance for a simple installation

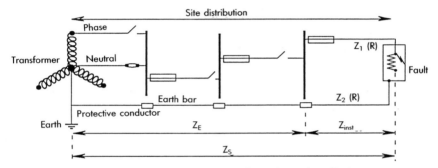

Fig. 12.1b Phase earth loop impedance for a large installation

(8) The actual value of Z_S for the circuit can be used to determine that the protective conductor has sufficient thermal capacity, by using the same formula given in (6), or by using Table 54G.

When using the formula given in (6) the fault current I_f is obtained from U_0/Z_s where U_0 is the voltage to earth. Time t is obtained from the protective device characteristic for fault current I_f. The factor k is obtained from the tables in Chapter 54 of the IEE Wiring Regulations and corresponds to the temperature of the protective conductor at the start of the fault for the type of conductor insulation being used.

Where the protective conductor is contained within a cable or grouped with live conductors, its temperature is taken as being the same as the live conductors. Where the protective conductor is the armour or sheath of a cable its initial temperature is taken as being 60°C for standard PVC insulation and 80°C for XLPE insulation.

Where a distribution board is to supply both fixed equipment having a disconnection time of 5 s and socket-outlets or circuits supplying hand-held equipment, there are two ways of providing suitable protection. The first is to ensure that the protective conductor from the distribution board back to the main earth bar complies with Table 41C. The second is to install another earth bar at the distribution board, installing equipotential bonding conductors from this earth bar to the same extraneous conductive parts to which the main equipotential bonding conductors are connected. The size of the equipotential bonding conductors is then the same as the main equipotential bonding conductors.

In many cases a final circuit will be one of a number connected to a distribution board which, in turn, is supplied by a submains cable. Under these circumstances the design currents of all the circuits must be summated, taking into account diversity, and the above procedure followed to select the size of the submain and the appropriate protective device at the source.

In lighting and socket-outlet installations, the various cables are usually contained in a single protective enclosure such as conduit or trunking and, therefore, there will be a greater build-up of heat than for a single circuit. This is avoided by using the grouping factors referred to in Chapter 8. However, there is a requirement in the regulations that cables can be easily drawn in or out, so the number of cables enclosed in a particular size of conduit or trunking has to be limited.

The IEE Guide to Selection and Erection gives tables for working out the maximum number of cables that can be contained in enclosures. The following is an example of the method used.

Example

Cables to be installed include $19 \times 2.5\,\text{mm}^2$, $21 \times 6.00\,\text{mm}^2$ and $9 \times 10\,\text{mm}^2$ single-core PVC-insulated stranded cables. Determine the size of trunking required.

Working

From Table A5 the required cable factors are:
for 2.5 mm   11.4
for 6.00 mm   22.9
for 10.00 mm   36.3
Summation $(19 \times 11.4) + (21 \times 22.9 + (9 \times 36.3)$
                    216.6 +   480.9 + 326.7 = 1024.2

From Table A6, the nearest trunking factor above 1024.2 is 1037 and the required size is, therefore, $50 \times 50\,\text{mm}$. Note that, as the difference between the two factors is 12.8, the space capacity only allows for one additional

2.5 mm cable and, consequently, it is preferable to use either 75×37.5 (factor 1146) or 75×50 (1555) to provide sufficient spare capacity for future extensions, providing the existing cables have already been derated to allow for further cables to be added to the group at a later date. If the cables installed have not been sized to allow for a larger number of cables being grouped together at a later date, then no additional circuits can be installed in the trunking no matter how much space there is, unless the additional currents all comply with the 30% lightly loaded cable factor explained in Chapter 8.

From the trunking, assume that six of the 2.5 mm^2 cables are to branch off in a 6 metre conduit run containing two bends. Determine the conduit size.

From Table A3, the cable factor is 30: total $6 \times 30 = 180$.

Refer to Table A4, first column, for length of run and read across to conduit factors under 'Two bends' to obtain nearest above 180. This indicates that 20 mm conduit (factor 182) is adequate but, again, with no allowance for future additions.

It must be noted that there are differences between the factors depending upon the circumstances, and it is essential that the correct ones are chosen.

Lighting

Although many lighting designs are within the capability of the installation designer, the subject is extensive and the reader is advised to study the numerous technical documents issued by such organizations as the Chartered Institution of Building Services Engineers (CIBSE) and a number of luminaire manufacturers.

One of the main considerations is that a lighting installation must provide illumination where it is required in the most effective and economical way. Economy should extend to both installation and running costs as lighting is, very often, one of the heaviest and most continuous loads.

It is also essential ensure that, in the event of a loss of power, as little disruption as possible is caused to the whole installation by splitting it into convenient groups and, if a three-phase supply is provided, balancing the load over the different phases. Additionally, in commercial and industrial premises which are covered by the Health and Safety at Work etc. Act and the Factory Regulations, emergency lighting (see BS 5266) is required on all escape routes and may also be necessary to allow dangerous processes to be safely shut down. Wiring for emergency lighting circuits must be segregated from other circuits unless the self-contained type of luminaire is installed, in which case a failure of supply to the luminaires will simply bring them into operation. (See also Chapter 15.)

A further consideration is that, in many enclosed luminaires, high temperatures are created and, consequently, final connections must be effected with suitable heat resistant tails such as glass sleeving.

Regarding circuit wiring, it must be emphasized that the running current of luminaires is not necessarily the major consideration especially where discharge fittings are used, and the designer must allow in the cable sizing for the effects of the associated control gear and harmonics. These can have a considerable effect on the power factor, which increases the current, and harmonics in some cases may be as high as one-third of the load current; on three-phase circuits, zero-phase sequence currents are in synchronism and, therefore, the neutral current may equal the single-phase current which prevents reduction of the size of the neutral. Section 6.3 of the Guide to Selection and Erection requires that, unless detailed information is obtainable from luminaire manufacturers, the volt-amperes of the fitting must be multiplied by a factor of 1.8 to obtain the circuit current.

Finally, as lighting installations are classified as fixed systems, the disconnection period under earth-fault conditions is 5 s for installations within the equipotential zone and 0.4 s for installations outside the zone. Therefore, Table 41B1, 41B2 and 41D are applicable for the maximum impedances.

Socket-outlets

As the IEE Wiring Regulations acknowledge the fact that socket-outlet circuits may be used to provide supply to either fixed equipment or portable apparatus, the disconnection period may be 5 s for the first or 0.4 s for the second within the equipotential zone. Before the designer decides which to use, however, he must ensure beyond any shadow of doubt that the former cannot be used for hand-held equipment. A further requirement, possibly more essential for domestic installations, is that any socket outlet that can be used for equipment outdoors, such as garden tools, residual current protection must be provided (Figs 12.2 and 12.3).

Socket-outlet circuits can be fed from either ring or radial circuits with some restrictions on the number of outlets per circuit. In both types of circuit spurs may be taken from the main system with limited restriction on the number of outlets provided that the spur is fused at the tee-off

Fig. 12.2 RCCB used for portable tools. (Courtesy of Power Breaker Ltd.)

Fig. 12.3 RCCB-protected 13 A socket-outlet for permanent installations in domestic or commercial premises. (Courtesy of Power Breaker Ltd.)

point. Standard ring final circuits must be fused at no more than 32 A which, of course, ensures that the total load is limited and, therefore, particularly in commercial and industrial premises, the designer must ascertain the possible usage of the outlets to avoid constant overloading and nuisance tripping. (See Chapter 8 for a worked example.)

Power equipment

As the requirements of power equipment are larger than those of lighting and socket-outlets so far as individual loads are concerned, each circuit must be treated individually and the same considerations of fault level and earth protection given to each. It is, however, also necessary to apply further precautions such as means of isolation for emergency, maintenance or functional purposes as detailed in Chapter 46 and Sections 476 and 537, and the starting requirements for motors. For the wiring of motor circuits, it is permitted to consider only the full-load currents of the machines but as, for direct-on-line (DOL) starting of squirrel-cage motors, the starting current may be up to seven times the full-load current, the resulting drop in voltage at the time of switch-on may be sufficient to prevent starting and, consequently, heavier cable may be required to reduce the voltage drop. To avoid this problem, other means of starting may be employed such as star-delta, auto-transformer (assisted starts) or soft-start but, in many cases, the choice is not that of the installation designer and, therefore, he must design accordingly.

Design schedule 1: basic design stages

The following are the main steps required to conform to the 16th Edition of the IEE Wiring Regulations:

(1) Obtain by calculation or by estimate the design load current I_b.
(2) Decide the type and rating, I_n, of protection.
(3) Calculate the current rating required for the cable I_t.
(4) Choose suitable cable size from standard ratings, I_{tab}.
(5) Calculate voltage drop.
(6) Check short-circuit capacity of cable against the protection capacity.
(7) Determine the maximum earth loop impedance Z_s allowed for the protective device being used.
(8) Calculate the ohms per metre of the circuit protective conductor, and choose from standard ratings. (Note minimum size restrictions.)
(9) Calculate value of maximum earth fault current.
(10) Check for shock constraint, i.e. disconnection period of protection.

(11) Check for thermal constraint of CPC using the adiabatic equation or for compliance with Table 54G.

The same procedures have to be followed with regard to submains cabling but, in this case, the voltage drop on the worst-case final circuit must be taken into account. Diversity between final circuits may be considered but the submain cable rating must be at least equal to the anticipated full load current of the distribution board.

Design schedule 2: relationship of current ratings

Formula $= I_b \leqslant I_n \leqslant I_z$ for overload protection
(Reg. 433–02–01)
Aide memoire – taken alphabetically, B comes before N and N comes before Z. Therefore order is I_b, I_n, I_z.

Circuit design

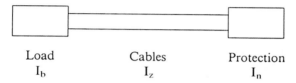

Load   Cables   Protection
I_b I_z I_n

(1) Load must be determined $= I_b$.
(2) Cables must carry load therefore load must be equal to or less than cable rating, $I_b \leqslant I_z$.
(3) The load must equal or be less than the protection therefore $I_b \leqslant I_n$.
(4) Result:

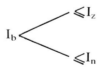

(5) BUT the protection must be equal to or less than the cable rating therefore $I_n \leqslant I_z$.
(6) From (2) and (5) we obtain $I_b \leqslant I_n \leqslant I_z$ and this relationship must always exist for overload protection. In all other cases $I_b \leqslant I_n$ and $I_b \leqslant I_z$, but I_z can be $< I_n$.

Design schedule 3: voltage drop

Having determined the size of live conductors to comply with the overload regulations, taking into account any correction factors for grouping, ambient temperature etc., the next step is to check that the circuit complies with voltage drop.

The installation is deemed to comply with the regulation for voltage drop if:

- The supply to the site complies with the Electricity Supply Regulations 1988 (as amended) and,
- The voltage drop from the origin of the installation up to the end of each circuit does not exceed 4% of the nominal voltage of the supply.

Each of the cable current-carrying capacity tables in Appendix 4 of the IEE Wiring Regulations contains the voltage drop per ampere per metre for the method used to install the cable. The voltage drop given in each table is based on the operating temperature of the conductor given in that table. The voltage drop values in the tables of Appendix 4 are given in millivolts per ampere per metre. For conductors up to $16\,mm^2$ only resistances are given while for larger conductors both reactances and impedances are also tabulated. The values in the tables include the phase and neutral conductors for single-phase supplies and the line voltage drop for three-phase supplies.

The first step is to use the impedance value to check that the voltage drop does not exceed the 4% allowed, i.e., 9.6 V for a 240 V supply and 16.6 V for a 415 V three-phase supply. If the voltage drop exceeds the value allowed, the next step is to look at the circuit to determine:

(1) Whether the circuit conductors are carrying their full rated current.
(2) Whether the circuit is feeding a load that has a power factor.

If either (1) or (2) are applicable corrections can be made to the voltage drop by using the formulae given in Chapter 8.

Design schedule 4: short-circuit protection

Having sized the live conductors a check must now be made to ensure the conductors are protected against short-circuit current.

If the breaking capacity of the protective device equals or exceeds the prospective short-circuit current at the point the device is installed, and the protective device for the circuit is also a fuse or a current limiting circuit breaker providing overload protection, it can be assumed that the conductors on the load side of the device are protected against short-circuit current and no calculations are necessary.

If the protective device is only giving short-circuit protection, for example the circuits feeds a motor, then a calculation is needed to ensure the conductors are protected against short-circuit current. The calculation is carried out in accordance with Chapter 8 using the following formula:

$$t = \frac{k^2 S^2}{I^2}$$

To ensure the conductors are protected the minimum short-circuit current is required. For three-phase and neutral circuits or single-phase circuits the minimum prospective short-circuit current is obtained from:

$$I_{pn} = \frac{V_{ph}}{Z_{pn}}$$

where V_{ph} is the phase to neutral voltage, Z_{pn} is the phase and neutral impedance from the source of supply to the point under consideration.

For three-phase three-wire circuits the minimum short-circuit current will occur when the fault is between two phases, in this case the short-circuit current is obtained from:

$$I_{pp} = \frac{V_L}{2Z_p}$$

where V_L is the line voltage and Z_p is the impedance of one phase from the source of supply to the point under consideration.

The value k is the maximum thermal capacity of the conductor and is obtained from Table 43A in the Regulations; typical values being 115 for general purpose PVC cable with copper conductors and 143 for XLPE cable with copper conductors.

Substituting the above values obtained into the formula will give the maximum time the fault current can be allowed to flow before the conductor is damaged or destroyed. The actual time taken for the protective device to disconnect the circuit is obtained by plotting the short-circuit current on the time/current characteristic for the protective device. If the actual time taken to disconnect the circuit is 0.1 s or less then the I^2t in the equation has to be obtained from the manufacturer by asking for the I^2t energy let through the device.

For checking that switchgear, isolators and motor starters etc., will withstand the short-circuit current flowing through them, the maximum peak fault current is required. This entails doing the calculations again, but this time with the resistance of conductors being taken at the minimum temperature at which the conductors could be operating when a fault occurs.

Since the maximum fault current is required the formula to use for three-phase three-wire or three-phase four-wire circuits is:

$$I_p = \frac{V_{ph}}{Z_p}$$

where V_{ph} is the phase voltage and Z_p is the impedance of only one phase, from the source of supply up to the point of the fault or the point at which it is required to know the maximum short-circuit current. Where the supply is only single phase then Z_p in the above formula is replaced by Z_{pn}.

Design schedule 5: shock constraint determination

Having determined the conductor sizes the circuit can now be checked to ensure protection against indirect contact. Checking protection against indirect contact entails the calculation of the maximum earth fault loop impedance, to ensure that it does not exceed the maximum allowed, for the type of circuit and protective device being used.

For voltages to earth between 220 V and 277 V in normal environmental conditions, the disconnection time for socket-outlet circuits or circuits supplying portable equipment intended for manual movement or hand-held equipment must not exceed 0.4 s. The disconnection time for fixed equipment within the equipotential zone can be 5 s.

The factors required for this exercise are:

Z_E impedance up to the circuit being designed.
Z_1 impedance of the circuit's phase conductor.
Z_2 impedance of the circuit's protective conductor.

If the sum of $Z_1 + Z_2$ is called Z_{inst} then $Z_S = Z_E + Z_{inst}$ which is a convenient formula to remember.

The advantage of using the above formula is its simplicity; knowing any two of the items enables the third to be obtained.

$$Z_S = Z_E + Z_{inst} \tag{12.1}$$
$$Z_E = Z_S - Z_{inst} \tag{12.2}$$
$$Z_{inst} = Z_S - Z_E \tag{12.3}$$

This arrangement is helpful if a circuit has to be extended. Knowing the maximum Z_S allowed for the circuit, and the earth loop impedance Z_E up to the start of the circuit, enables Z_{inst} to be found using Equation (12.3). Z_{inst} can now be divided by the impedance per metre for the cable, i.e., $(Z_1 + Z_2)$ per metre. This will give the total length allowed for the circuit for protection against indirect contact.

When calculating Z_S the resistance of the conductors is taken at the average of the operating temperature for the conductor and the limit temperature of the conductor's insulation. The calculated value of Z_S is then compared with the maximum permitted Z_S given in the tables for the type and rating of the protective device being used to protect the circuit.

Design schedule 6: protective conductor suitability

Protective conductors have to be checked to ensure they have sufficient thermal capacity under fault conditions. Two methods are available, using Table 54G or by calculation. In practice, except for the smallest jobs, the calculation method will give the most economical result.

The calculation method uses the same formula as used for checking short-circuit protection, but re-arranged so that the size of the protective conductor is unknown, i.e.,

$$S = \frac{\sqrt{I_f^2 t}}{k}$$

The value of Z_S determined for indirect contact can be used to calculate the phase-to-earth fault current I_f as explained in Chapter 8.

The calculation gives the minimum cross-sectional area allowed for the protective conductor. This is then compared with the actual area of the protective conductor used, which must not be less than the calculated value.

If the disconnection time t is 0.1 s or less then $I^2 t$ is obtained from the manufacturer of the protective device, as in the case of short-circuit current calculations.

If the phase conductor has been sized due to the prospective short-circuit current, and I_f is less than the prospective short-circuit current, the protective conductor must be checked by calculation and the phase conductors will also need rechecking that they are still protected with the lower fault current.

Except for special situations a protective conductor must be installed for each circuit working at low voltage. This conductor is referred to as a circuit protective conductor (CPC).

If the protective conductor is not part of a cable or formed by conduit, ducting, trunking or not contained in an enclosure of a wiring system, then the minimum size allowed is a 2.5 mm^2 copper conductor.

Where the circuit protective conductor is common to several circuits its cross-section area should be either: calculated by using the most onerous values of fault current and disconnection time, or determined by using the largest phase conductor in conjunction with Table 54G.

Design schedule 7: equipotential bonding

The definition of equipotential bonding is: 'Electrical connection maintaining various exposed-conductive parts and extraneous-conductive parts at substantially the same potential'.

To try and create an equipotential zone main equipotential bonding conductors are installed from the main earth bar to the main extraneous conductive parts that run throughout the building, including any metal parts that pass out of the building into the ground. The equipotential zone is not complete since a fault to earth within the equipotential zone will result in a voltage appearing between the exposed conductive parts of the faulty circuit and any extraneous conductive parts, even though the exposed and extraneous conductive parts are both connected to the main earth bar.

The parts included in the bonding are: hot and cold water and other service pipes, gas installation pipes, central heating and air conditioning systems, exposed structural metal parts and the lightning conductor system.

Where the supply is protective multiple earthing (PME) the bonding conductor size is determined from Table 54H in the regulations. Where the supply is not PME the size of the main equipotential bonding conductors shall be half the size of the main earthing conductor, subject to a minimum size of $6\,mm^2$; the maximum need not exceed $25\,mm^2$ if the conductor is copper or an equivalent conductance in other metals.

Where there is an insulating break in any of the services a supplementary bond must be installed to maintain continuity; it follows that this bond should have the same cross-sectional area as the main equipotential bonding conductors.

Where an extraneous conductive part that requires bonding is remote from the main earth bar, an extraneous conductive part that is already connected by bonding conductors can be used, providing it is neither a gas or oil service pipe.

Design schedule 8: supplementary bonding

Special installations, such as bathrooms, require additional bonding to be carried out irrespective of the main equipotential bonding and minimum sizes are given in Regulation 547−03. In the case of bathrooms or shower rooms all extraneous and exposed conductive parts must be bonded together. Where an exposed conductive part is connected via a flex outlet, the protective conductor within the flexible cable can also be used as the bonding conductor, the bonding conductor terminating at the earth terminal in the flex outlet.

Where a circuit cannot be disconnected within the time specified (either 0.4 or 5 s) either local supplementary bonding has to be carried out or the circuit has to be protected by a residual current device. Where supplementary bonding is used to bond between simultaneously accessible exposed conductive parts and extraneous conductive parts, the resistance of the bonding conductor can be determined by the following formula:

$$R \leqslant \frac{50}{I_a}$$

where I_a is the current that disconnects the overcurrent device within 5 s or for a residual current device is $I_{\Delta n}$. The figure of 50 V is reduced to 25 V in some of the special installations and locations covered by Part 6 of the IEE Wiring Regulations.

The supplementary bonding conductor is then subject to the minimum sizes given in Regulation 547−03 below. The following are based on the supplementary bonding conductor being sheathed or otherwise mechanically protected. Where this is not the case the minimum size allowed is 4 mm^2.

Regulation 547−03−01

The conductance must be not less than that of the smallest protective conductor when two exposed conductive parts are connected together.

Regulation 547−03−02

If connecting an exposed conductive part to an extraneous conductive part, it shall have a conductance not less than half that of the protective conductor connected to the exposed conductive part.

Regulation 547−03−03

For the bonding of two extraneous conductive parts the minimum cross-sectional area is 2.5 mm^2 if the conductor is sheathed or otherwise mechanically protected and 4 mm^2 if not. Where one of the extraneous conductive parts is connected to an exposed conductive part then Regulation 547−03−02 applies to all the supplementary bonding conductors.

Supplementary bonding can be provided by:

(1) A supplementary bonding conductor.
(2) A conductive part of a permanent or reliable nature.
(3) A combination of both (1) and (2).

Since a conductive part can be used for bonding, the steel frame of the building could be used as a supplementary bonding conductor.

Additional supplementary bonding has to be installed where called for in Part 6 of the IEE Wiring Regulations, which covers areas where extra precautions have to be taken to protect against electric shock. One such area, bathrooms, has already been mentioned above. Another area is swimming pools, where an earthed metallic screen has to be installed in the floor when a solid floor is used and, except for SELV (was safety extra-low voltage in the IEE Wiring Regulations 15th edition) circuits, supplementary equipotential bonding has to be carried out in zones A, B and C between extraneous conductive parts and the protective conductors of all exposed conductive parts within these zones.

Design schedule 9: earthing

Every installation has to be provided with a main earthing terminal to which are connected the circuit protective conductors, main equipotential bonding conductors, any functional earthing conductors and the bonding conductor to the lightning protection system.

The main earthing conductor has to be sized in accordance with Section 543 and in addition when buried in the ground the earthing conductor is subject to the minimum sizes specified in Table 54A. Earthing is covered in more detail in Chapter 16.

Examples of protecting against indirect contact

To enable the calculations suitable for the commercial and industrial installation to be illustrated the following information has been obtained from the *Handbook on the IEE Wiring Regulations* by T.E. Marks.

200 kVA transformer impedance of each winding 0.014 + j0.0423
Impedance of 150 mm^2 4-core PVCSWA & PVC copper cable per
    1000 m
Phase conductor 115°C = 0.171 + j0.076 for each core and
Phase earth loop impedance = 1.04 Ω per 1000 m.
Impedance of armour = 0.856 Ω per 1000 m.
Impedance of 95 mm^2 4-core PVCSAW & PVC copper cable per
    1000 m

Phase conductor 115°C = 0.266 + j0.077 for each core and
Phase earth loop impedance = 1.52 Ω per 1000 m.
Impedance of armour = 1.246 Ω per 1000 m
Impedance of 16 mm^2 copper cable at 115°C = 1.587 Ω per 1000 m.
Impedance of 50 mm × 50 mm steel trunking is 0.00345 Ω per metre.
Resistance of 2.5 mm^2 PVC insulated copper cable = 10.226 Ω per 1000 m.
Resistance of 1.5 mm^2 PVC insulated copper cable = 16.698 Ω per 1000 m.
Maximum value of Z_s for a 40 A HRC fuse = 0.86 Ω for 0.4 s and 1.41 Ω for 5 s.
Maximum value of Z_s for a 20 A rewirable fuse is 1.85 Ω for 0.4 s and 4.0 Ω for 5 s.

Example 1

A factory distribution scheme, illustrated in Fig. 12.4, takes its supply from a 200 kVA transformer which feeds a main switchboard by a 10 m length of 150 mm^2 PVCSWA & PVC cable. The main switchboard feeds a final distribution board with a 100 m length of 95 mm^2 PVCSWA & PVC cable. A circuit taken from the final distribution board feeds some three phase socket-outlets, the size of cable used is 16 mm^2 copper and they are installed along with other circuits in a 50 mm × 50 mm steel trunking. If the length of the final circuit is 75 m and the protective device is a 40 A HRC fuse determine whether the circuit complies with the requirements for indirect contact.

Working

The worst possible case for protection against indirect contact would be if the cables back to the source transformer attained the 115°C temperature for PVC conductors under fault conditions i.e., the average of the permitted operating temperature and the limit temperature of the conductor's insulation.

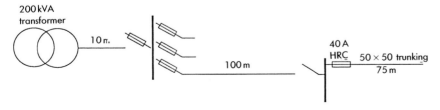

200 kVA
transformer

10 m.

100 m

40 A
HRC
50 × 50 trunking
75 m

Fig. 12.4 Factory distribution

The earth loop impedance is worked out as follows:

Transformer	0.014 + j0.0423
Cable to main switchgear $(0.171 + j0.076) \div 100$	0.00171 + j0.00076
Switchgear to distribution board	
$(0.266 + j0.077) \div 10$	0.0226 + j0.0077
Final circuit phase cable $(1.587 \times 75 \div 1000)$	0.119
Armour of $150\,mm^2$ cable $0.856 \div 100$	0.00856
Armour of $95\,mm^2$ cable $1.246 \div 10$	0.1246
Impedance of trunking 0.00345×75	0.25875
	0.54922 + j0.05076

The value of Z_s for the final circuit is:

$$Z_s = \sqrt{0.54922^2 + 0.05076^2} = 0.55\,\Omega$$

The value of Z_s calculated is now compared with the maximum Z_s allowed for the circuit which, for a 40 A HRC fuse, is $0.86\,\Omega$ for a 0.4 s disconnection time (since they are socket-outlets). The circuit being acceptable for protection against indirect contact.

The method used for calculating Z_s illustrates how resistances and reactances for cables and transformers are dealt with by keeping the resistances and reactances separate until the impedance is required.

The above calculation is not strictly correct, since for the last three items impedances have been treated as resistances, but in practice (unless large cables and transformers are being dealt with) the phase earth loop impedance of each cable can be added together. This does, however, give a more pessimistic result and errs on the safe side.

Example 2

An outside light is to be installed on a house using 2.5 PVC twin and CPC cable. If Z_E external to the distribution board is $0.35\,\Omega$ and the circuit is to be protected by a 20 A rewirable fuse, what can the maximum length of the circuit be to satisfy protection against indirect contact?

Working

The lighting fitting will be outside the equipotential zone so the disconnection time allowed is 0.4 s, it is fixed equipment and does not, therefore, require protecting by an RCD.

Design schedule 5 can be used to determine how much impedance is allowed for the cable.

Using Equation (12.3)

$$Z_{inst} = Z_s - Z_E$$

The maximum value of Z_s allowed is $1.85\,\Omega$ therefore

$$Z_{inst} = 1.85 - 0.35 = 1.5\,\Omega$$

This is the maximum impedance allowed for the phase conductor R_1 and the protective conductor R_2.

The impedance per metre of 2.5/1.5 cable is $(10.226 + 16.698)/1000$ at 115°C

$$\text{Maximum length of cable} = \frac{Z_{inst} \text{ allowed}}{\text{impedance per metre for cable}}$$

$$= \frac{1.5 \times 1000}{(10.226 + 16.698)} = 55.71\,\text{m}$$

Thus the circuit cable can be made 55.71 m long, it may not be suitable at this length for short circuit protection, size of CPC and voltage drop, but the circuit will be disconnected in 0.4 s.

Chapter 13
Special Cabling Requirements

Although PVC-insulated cables are suitable for most of the general wiring requirements in domestic, commercial and industrial situations, circumstances may dictate, either through technical necessity or statutory demands, that further precautions are necessary to prevent the possibility of danger or to give increased security, as described below.

Lighting

The two main areas of concern are related to heat build-up in luminaires and surges created by discharge lighting. In totally enclosed luminaires, high temperatures may arise due to the lack of ventilation and, while many manufacturers provide for this by supplying fittings complete with high-temperature tails or heat shields in compliance with BS 4533, this is not always the case with the cheaper or some imported fittings and the responsibility lies with the designer/installer to ensure that wiring in proximity to the fittings is suitable (Fig. 13.1).

Reference is made elsewhere to the fact that discharge-type fittings may

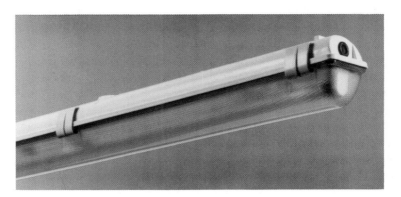

Fig. 13.1 Thorn Lighting Ltd LUV luminaire for hostile environments where a jet-proof, dust-tight fitting is required (courtesy of Hugh King, Thorn EMI). Note that the nature of the enclosure prevents natural ventilation and, therefore, necessitates high temperature internal wiring, etc.

entail the use of higher current rated cables to avoid unnecessary temperature rises, but the effects of high discharge currents during switching operations may have more drastic effects by causing a cable to disintegrate completely, particularly in the case of MICC cables. These are susceptible to current and voltage surges which may be avoided by the use of current-limiting devices obtainable from the cable manufacturers. Regulation 422−01−02 stipulates the methods of protection to be adopted, where electrical equipment in normal operation has a surface temperature sufficient to cause a risk of fire.

Emergency lighting

During recent years, the importance of emergency lighting has been greatly emphasized by the Health and Safety at Work etc. Act and the Code of Practice BS 5266: Part 1 first issued in 1975. The Act is directed mainly at factories, although it is undergoing revisions which will make it equally applicable to other premises such as hospitals, theatres, hotels, offices and shops, while BS 5266 covers practically all types of premises excluding houses, cinemas and certain specified places of entertainment. CP 1007: 1955 is applicable to cinemas only.

Generally, the cable installation for an emergency lighting system should comply with the IEE Wiring Regulations but care must be taken to ensure that all wiring possesses inherently high resistance to attack by fire and adequate mechanical strength. This allows the use of various standard types of cable, provided that suitable means of protection are employed.

One main requirement, referred to briefly in the Regulations under the subject of segregation, is that when emergency luminaires are supplied from a remote source the wiring system must be mechanically separated from other systems by rigid and continuous partitions of noncombustible material. Consequently, multicompartment enclosures are suitable, also MICC cables without further precautions. Segregation is not a requirement when self-contained luminaires are installed, as a failure of the supply will only cause them to operate (Fig. 13.2).

Precautions to be taken at the source of supply for an emergency lighting system are that cables between the source and a battery charger combination should be a fixed installation, which precludes plugs and sockets, while those cables from the battery to a protective device, i.e. the load circuit cables, must be separated from each other and not enclosed within metal conduit, ducting or trunking. Segregation must also be applied between the d.c. and any a.c. cables.

Fig. 13.2 Examples of the range of emergency lighting self-contained luminaires available for commercial and industrial applications (courtesy of Chloride Bardic Ltd).

Fire alarms and detection

Many of the requirements in the previous section were the results of experience gained from CP 1019 which was the original Code of Practice for fire alarm and detection systems, now superseded by BS 5839. Consequently, if the same principles regarding mechanical protection, high fire resistance and segregation, etc. are applied, little further can be added.

Perhaps an additional precaution, which is more applicable to this type of installation to avoid false alarms rather than to improve safety, is to ensure that where the high frequency circuits are installed adequate screening is applied between the different circuits.

In many of the present systems, advantage is being taken of multiplexing to reduce the number of cables in extensive installations or, as for control and instrumentation, optical fibres. However, some fire prevention authorities may still not be prepared to approve such alternative wiring systems.

Power systems

Some of the problems arising in the installation of power cables have been referred to in Chapter 8, i.e. high or low ambient temperatures, grouping, thermal insulation, type of protective device employed and voltage drop considerations. Under normal circumstances, correctly chosen protective devices are adequate to deal with disruptions such as overloads, short-circuits and earth-faults on low voltage systems but, on high voltage networks, transients may occur which create high stresses on cable insulation and, therefore, it may be advisable to install screened cables which have the effect of grading such stresses between cores or between cores and earth.

The handling and installation of all types of cable is an area which does not always receive the necessary attention; some PVC cables, for instance, should not be installed during temperatures below 0°C as flexing will damage the insulation, while high temperatures will soften the PVC, causing it to strip if pulled into conduit, ducting, etc. Damage may also be caused to cables by drawing them into rough-edged enclosures, e.g. burred conduits, over stony surfaces or bending them tighter than the recommended radii indicated in Appendix I of the Guide to Selection and Erection of Equipment. Large armoured cables are impressively strong, but even these, when being drawn into ducts, may be damaged if the correct type of grip-sleeve (or sock) and hauling equipment is not used, as too high a torque may stretch the cable cores or strip off the insulation and sheathing.

Particularly with the smaller armoured cables, if armouring is to be used as the protective conductor, the impedance must be checked to ensure that it complies with the IEE Wiring Regulations, otherwise additional conductive material must be incorporated in the protective circuit.

Control and instrumentation

Modern systems for control and instrumentation utilize electronic means (rather than power circuitry) which are more likely to be affected by low voltage systems, and precautions such as segregation and screening must be employed. As stated in Chapter 7, cables are available to suit all types of system but, as requirements vary between manufacturers of electronic equipment, advice should be sought at an early stage. Chapter 7 also referred to the increasing use of multiplex systems and fibre-optics cables which simplify installation work by reducing the number of cores required for the most complex systems and, in the case of the latter, eliminate completely the possibility of interference from other circuits.

Hazardous areas

Danger in a hazardous area arises initially from the type of materials being processed rather than from the electrical installation, but a great degree of responsibility rests upon the designer to ensure that the installation does not contribute to the hazard by the introduction of flammable materials, high surface temperatures, arcs or sparks to the atmosphere. For these reasons, every care must be taken to avoid the overloading of cables or the inclusion of sheathing materials which easily burn and give off toxic gases.

Hazardous areas are defined as areas in which there is a potential hazard due to the presence of flammable gases or vapours, and these are given zonal classifications according to the degree of danger as follows:

Zone 0   Mainly underground mining operations.
Zone 1   An area in which flammable gases or vapours exist in the atmosphere continuously, for long periods or during normal process operations.
Zone 2   Generally, areas adjacent to zone 1 in which it is possible for flammable gases or vapours to be present for short periods, e.g. on the occurrence of an unforeseen incident.

For more detailed guidance, reference should be made to BS 5345: Part 2, Classification of Hazardous Areas.

From the above it will be appreciated that different degrees of hazard are recognized and, consequently, these affect the type of electrical installation, particularly with regard to equipment. It is essential, therefore, to ascertain which zone is applicable before commencing the electrical design, this information generally being available from the process plant user or the Health and Safety Executive (in the UK).

The types of cable installation differ little from those in general use and include: firstly, single or multicore cables in heavy gauge solid or welded conduit; secondly, sheathed armoured cables; and finally, mineral-insulated cables. The sheaths of unenclosed cables must, of course, be impervious to the possible effects of the gases or vapours present, be heat resistant and nontoxic. Termination of cable installations should be such that it is impossible for gases or vapours to travel through the enclosures or the interstices of the cables, particularly between zones, but provision for this is incorporated in the terminating equipment by the manufacturers. All cable boxes, glands, etc. for use in hazardous areas are subjected to stringent tests by the appropriate national test houses and are certified for use in the various zones. Note that the higher the zone number, the lower is the possible hazard and, while it is permitted to install equipment suitable for a more hazardous area in one that has a lower potential danger, the reverse is not the case.

It is strongly recommended that hazardous area installations be kept to the absolute minimum and, therefore, cabling not directly associated with plant in such areas should, whenever possible, be routed elsewhere. Extreme care should also be taken where cables pass from one area to another of different type to ensure that barrier boxes are installed in the cable runs adjacent to but outside the more dangerous zone, and that cable apertures are adequately sealed with fire-resistant materials.

Full details of the requirements are included in CP 1003 (for older installations), BS 5345 and BS 5501, all of which are available from the British Standards Institution or may be referred to in many major libraries.

Chapter 14

Computer Control of Environmental Services

Building users require services to meet the environmental and functional needs associated with a particular type of building, and these services vary considerably according to the type of building involved. However, the basic requirements are for comfort, safety, security and operational utilities.

General building classifications are residential, commercial, industrial and public, and the requirements differ according to the particular purpose of the building. The complexities of the services also relate to the requirements and additionally to the size and class-type of the building.

Building management systems

Thermal comfort, lighting and electrical power are always required, and residential building represents this basic level of need.

The operation, control and economics of these systems on small projects can be handled by the individual but the management of building services systems for a larger establishment is more difficult due to the variety, increased complexity and lack of individual responsibility for the systems involved. The capital and operating costs of such systems are also of great concern to the occupier. The more complex control systems are termed building management systems (BMS). They are employed in commercial, public and industrial buildings and control the following services: heating, ventilation, air conditioning, steam, refrigeration, gas, water, general and emergency lighting, emergency electrical systems, power distribution, mechanical transportation, fire detection alarm and fighting systems, general and noxious fume ventilation, security and waste disposal.

The technical complexity and distribution of these systems throughout a building make control, management and maintenance a difficult proposition for staff and, as a result, many buildings do not function efficiently and economically. System design complexity has also increased due to the growth of energy conservation technology involving heat recovery, combined heat and power, and new environmental system designs.

Since the early 1970s, the influences of increasing energy costs and improved low cost microprocessor technology have revolutionized the methods of control and monitoring employed by building services engineers. The use of computers in the control and reporting functions is replacing traditional methods and is becoming an accepted part of large building design. Falling microprocessor costs now make these facilities available and relatively economical to much smaller buildings.

Previously, control of the services was by separate dedicated control systems with panels positioned local to the particular plant. In such localized systems, fault alarms and control adjustments had to be carried out at each location and required the attendance of engineers, necessitating routine visits to each plant in the building. On large sites containing several buildings, faults could remain unattended for considerable periods or until reported as complaints by the building occupants; for companies having properties scattered throughout the country, the problem is more severe and may involve long journeys before faults can be identified and rectified.

The advent of building management systems (BMS) now means that all services can be monitored and reset from a central location without delay or movement by the engineer. BMS can also advise on preventative maintenance schedules, thereby improving overall plant reliability and operating efficiency. Consequently, plant operation is greatly simplified by allowing an engineer to reset any control level, monitor energy consumption, organize maintenance and make fault diagnosis from a central location. Remedial action is quicker and can often be carried out by a smaller engineering staff than would be required otherwise. Figure 14.1 is a typical control console.

The advantages in improved performance and reduced energy and operating costs have to be equated to higher first cost.

BMS incorporates control and monitoring of all systems such as environmental, fire and security but, in some cases, separate dedicated fire protection systems are favoured by the authorities and reference to local fire codes and regulations is essential. The integration of fire protection and security systems into BMS is currently under debate, and the engineer should seek the guidance of the local fire prevention officer. It is possible to keep the fire protection system panel separate while still providing communication links with the BMS for alarm and reporting purposes.

Data transmission methods

The two main control options offered by BMS are centralized intelligence and distributed intelligence.

The centralized intelligence system carries out all the control and

Fig. 14.1 Typical microcomputer displaying data from remote stations. (Courtesy of Trend Controls Ltd.).

reporting functions by a single computer system and, although such systems are cheaper to install, they are vulnerable to complete breakdown if a failure occurs with the computer.

Back-up power supplies such as the UPS systems referred to in Chapter 3, although required for any BMS, need more consideration for centralized intelligence systems. A parallel systems structure and duplication of equipment to provide redundancy facilities may also be necessary, depending on the level of reliability required or the importance of the functions.

Distributed intelligence systems also have a central computer, with the addition of remote intelligent outstations capable of carrying out all control functions independent of the main computer. Outstations are located near to the building zone which they serve, as in the earlier systems, and are programmed to perform the required control functions. The outstation can, however, be interrogated and reset from the main computer and also communicate routine information and alarms as required by the plant operator (Fig. 14.2).

Although more expensive to install, most BMS are of the distributed intelligence type because of the improved reliability they offer, i.e. they are not totally dependent on the central control station. Each outstation

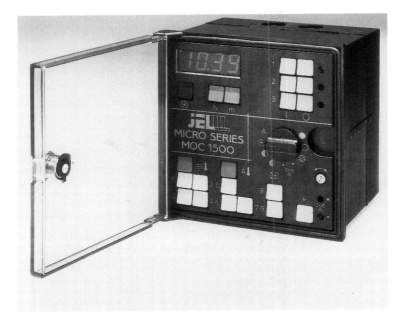

Fig. 14.2 Typical small low cost microprocessor-based controller. These provide a level of control and energy savings normally found only in expensive control systems.

can operate independently of the main computer and, therefore, provides more reliability. Failure of the computer does not interrupt the outstations, which can continue to carry out their control functions to the local plant.

The central computer communicates with the outstations through a standard interface and either a dedicated line, a leased line or a telephone switched line. Smaller systems on one site are usually connected using a twisted pair in either a ring, star or tree network.

Systems are being developed in which the control signals are multiplexed and transmitted around the building using the power distribution wiring. Special equipment is required to isolate the connections from the power at the transmitting and receiving points, thus increasing the possibility of failure, and as yet most control manufacturers employ separate wiring systems. The cost of multiplexing and power isolation equipment has to be considered against the cost of direct wiring, and in most cases direct wiring is cheaper and more reliable.

Optic fibre transmission is currently being installed and allows very fast transmission. As BMS becomes more widely accepted, it is likely that system capacities will increase to a level that will make optic fibre technology an attractive proposition.

For transmission distances within the building or site up to about 1.6 km, telephone lines are usually employed. The computer and outstations are connected to the telephone line via a modem, which converts the input signals to pulses which are transmittable on the telephone line, thus enabling information to be transmitted and received.

Two types of modem are currently in use: direct and autodial. The direct modems have leased lines permanently connected and are used on sites with lines limited as above between the computer and outstation. Greater distance almost invariably requires signal boosters to maintain satisfactory operation, as weak signals introduce errors.

In many cases, particularly for remote sites, the cost of a permanently connected leased line is not justified since the transmission of data and information occurs intermittently. Under these circumstances, an autodial modem restricts the use and cost of telephone lines by automatically communicating only when required. Autodial modems are therefore used for large or multiple sites, and the telephone line is accessed through the British Telecom system.

The cost of modems varies with quality, and it may be cost effective to install better modems to improve transmission accuracy and have less dependence on the quality of the transmission line. The choice of modem has to be determined taking into account the economics, actual site layout, transmission distances and the consequences of inaccuracy. Modems generally operate over a range of speeds (baud rate) with generally increased cost for the higher speed, e.g., 9600 baud is now common. (1 baud = 1 bit/second.) The baud rate is software set to suit the particular system, and some manufacturers use higher transmission rates for directly wired systems.

Autodial communication with the computer can be either direct dialling by the operator, programmed down-loading of data at predetermined times, or priority alarm reports.

It is advisable to have more than one line per station, one of which is dedicated to priority reporting and the other to routine reporting and monitoring. This allows each function to be effective without conflict with the other.

On large systems, the down-loading of data can take considerable time, is expensive in telephone time and is normally carried out at nonpeak periods. Computer access is sometimes restricted to once in every 24 hours to limit telephone charges, but can be accessed at any time if required by the plant operator.

Manufacturers are currently offering systems with over 200 outstations, each having many points, and the data associated with these is considerable.

Control monitoring stations

A typical control monitoring station consists of a microcomputer, visual display unit (VDU), backing store and printer. One station is usually installed in a central location, but systems with several stations working on a master/slave principle can be obtained for large sites (Fig. 14.3).

Buildings under phased development can be provided with a distributed intelligence system without a central computer. Under these circumstances, the outstation can be programmed and interrogated directly by portable hand-held microcomputers which are taken around the site by personnel and plugged into the outstation as required. A central computer can be added to the system at a later date, when the development is complete or finances allow.

The location of the central computer requires careful consideration and

Fig. 14.3 Typical central control and monitoring equipment includes printer, keyboard, VDU and transmitter/receiver.

should be provided with a clean power supply with back-up, as referred to previously. If connected to fire safety systems, it should be located where it can be easily accessed and interrogated by the fire brigade. The fire alarm protection system may be connected directly to the fire brigade, but the information supplied at the control station is likely to be more comprehensive and useful for directing fire-fighting operations and controlling services that may affect safety or the spread of fire.

Smoke removal and staircase pressurization systems for high rise buildings are often an integral part of the mechanical ventilation or air conditioning system. When this is the case, ready access by the fire brigade for system status and control is essential.

Microcomputers

Over recent years, the memory capacity of microcomputers has risen spectacularly, while the cost is continuing to fall. In these circumstances, the application of BMS can be expected to rise rapidly.

The clear implication to the building owner or operator is that the improved system performance, reliability and reduction in operating costs offered by BMS will provide a very attractive payback investment. At current cost levels, a system can be installed with sufficient capacity for extensive maintenance and energy conservation programmes quite economically.

The increased memory also allows more flexible and user-friendly software to be used, so that the operator is provided with easily understood graphic displays of the building and system under his control, rather than having to interpret a mass of data.

Some software designers provide graphics displays, which allow the operator to obtain colour drawings of any part of the building or system, with a zoom facility to enlarge those parts of the system of particular interest.

Maintenance schedules can alert operators to routine tasks and statistical programs permit data analysis for energy consumption, operating costs, etc. Such software is demanding on random access memory (RAM) capacity but RAM is now very cheap.

VDUs and keyboards

High resolution VDUs enable text and graphics to be displayed. Communication with the system software can be through keyboard, mouse or touch screen. The touch screen is the simplest but least flexible method and has not been widely adopted by BMS manufacturers. Mouse controllers are devices which are moved about a worktop to position a pointer on the VDU and, with suitable software, give quick access to facilities for operators who are not keyboard trained.

Most systems do, however, employ keyboards, as this provides the most flexible access and, as many software systems are menu driven, keyboard dexterity is not essential.

Backing store

Backing memory systems have also developed considerably.

Backing memory devices permanently store programs and data provided to and from the system. The devices are usually floppy disk or hard disk drives, but other alternatives are available.

Floppy disk drives come in various sizes and capacities, the most common being 3.25 in which is better protected against damage. Disk capacities are typically 1 MB, but for most large buildings requiring BMS this is insufficient. The major alternative is the hard disk which has much greater capacity and speed than can be obtained from 3.25 in floppy disks. Although Bernoulli have developed a removable floppy disk and disk drive with capacities at present up to 150 MB.

Hard disk drives have capacities typically over 100 MB. The mechanisms are more sensitive because of the fine tolerances between the read/write heads. It is essential to eliminate smoke and dust from the atmosphere, and vibration has to be avoided. Where the loss of data would be inconvenient, back-up memory stores using tape streamers or floppy disks should be installed.

Memory storage includes: write once, read many times (WORM) drives, re-writable optical drives, removable optical drives, removable hard disks, the Bernoulli removable floppy disk drive as well as hard disks. The advance in technology has been such that storage is no longer a problem.

Printers

The output of data and information from the system is transferred to paper copy by a printer. This is essential so that a readable log of the system performance is available for distribution to other interested members of the engineering and management team.

Since the quality of presentation is generally not important, dot matrix printers are normally suitable. Because the output is usually numerical data, in tabular or graphical form, it is advisable to obtain a printer with a wide carriage so that a larger number of columns can be accommodated. Dot matrix printers are relatively cheap, operate at higher speeds than the slower daisywheel printer and are, therefore, generally preferred.

Many present generation dot matrix printers can also operate in a letter quality mode at slower speed and have colour options available.

Laser printers offering both high speed and quality presentation are now available, but high cost restricts their use at present to situations where high volume reproduction is required.

Chapter 15

Emergency Lighting

Prior to the issue of BS 5266 in 1975 the requirements for emergency lighting in buildings were largely governed by the dictates of local authorities in the form of building regulations. Section 40 of the Factories Act, 1961 gives instructions on what precautions have to be taken to prevent fire and the provisions that have to be made in the factory in the event of a fire occurring, but gives no guidance on the provision of emergency lighting.

Part 1 of BS 5266: 1975 was completely revised in 1988 and the 1975 version was withdrawn. Although guidance on the provision of emergency lighting is available in the form of a British Standard, places of entertainment and cinemas still come under the control of the local authorities. The Cinematograph (Safety) Regulations 1955 require that adequate emergency lighting is provided to enable the public to see their way out of the cinema. At present emergency lighting in cinemas and theatres is covered by CP 1007: 1955, but these will ultimately be included in Part 2 of BS 5266 which will also give guidance on the emergency lighting in dance halls, ballrooms, licensed bingo premises and ten-pin bowling centres.

It is now recognized that an important function of emergency lighting is to provide a means of locating and operating fire alarm points and to provide sufficient illumination in open areas to reduce the risk of panic, in addition to its main function of illuminating the escape routes from the building.

Types of installation

There are two main types of emergency lighting: standby lighting and emergency escape lighting, the latter being further subdivided into:

- Escape route lighting.
- High-risk area lighting.
- Undefined escape routes (open area lighting) referred to as antipanic area lighting.

Escape route lighting has to provide illumination for, and clearly define, the escape route, it must also ensure that any fire alarm points or fire fighting equipment is illuminated. The escape route lighting has to operate when local areas lose their electricity supply in addition to operating when there is a complete failure of the mains supply.

The illumination called for in BS 5266: Part 1: 1988 for escape routes is for the illumination along the centre line of the escape route to be 0.2 lx with the additional requirement that for escape routes up to 2 m wide at least 50% of the width of the passage should be illuminated to 0.1 lx, wider escape routes being treated as multiples of 2 m wide escape routes.

The draft European standard calls for the illumination to be 1 lx, but the UK can still use the 0.2 lx lighting level providing there are no temporary or permanent obstructions along the escape route.

High-risk area lighting is for those areas using machinery or materials that could be a danger to personnel if the lighting failed, such as acid baths and conveyors etc. The amount of illumination to be provided should be sufficient to enable personnel to leave the area safely and may, in certain circumstances, have to be the same level of illumination as the main lighting. As such, a high risk area is not defined in BS 5266, but the draft European standard calls for at least 10% of the normal lighting or a minimum of 15 lx. This illumination has to be provided within 0.25 s of a normal supply failure. It should be noted that if the emergency lighting is provided by a generator it may fail to provide the lighting within 0.25 s and other arrangements will have to be made.

Where continuous operation is required, even if there is a failure of the main lighting, standby lighting should be provided. The amount of lighting provided will be dependent upon the degree of danger present and the degree to which instruments have to be read or adjustments made to the machinery and can involve providing the same level of illumination that the main lighting provides.

Antipanic or open area lighting is for those areas where no defined escape route can be indicated, such as in large shopping areas or open-plan offices etc. The level of illumination called for in BS 5266: Part 1: 1988 is an average of 1 lx over the whole area. The requirements of the draft European standard are similar, but calls for a minimum illumination of 0.5 lx anywhere on the floor, thus the lighting fittings will have to be arranged symmetrically to achieve this illumination and will require the provision of more lower wattage units rather than a few high output lighting fittings.

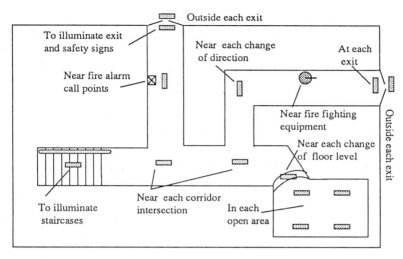

Fig. 15.1 Mandatory locations of essential escape lighting luminaires.

Location of escape route lighting

The design and positioning of the escape route lighting has to take into account that luminaires have to be installed internally and externally at each exit and emergency exit door to the premises and that a luminaire is required near each intersection or change in direction of a corridor, at each change of level and for each flight of stairs, near fire fighting equipment and fire alarm call points and to illuminate exit and safety signs, as illustrated in Fig. 15.1.

The British Standard defines 'near' as being within 2 m in a horizontal direction. Although lifts are not an escape route, emergency lighting should be installed in them. This is to provide comfort to those stuck in the lift and lessen the risk of panic where persons are in a confined space. Similarly escape route lighting should be provided for moving stairways and walkways. Toilets should also be illuminated unless they do not exceed 8 m² in floor area and are illuminated by borrowed light. The British Standard calls for motor generator rooms, control rooms, plant rooms, switch rooms and areas containing the equipment for the mains and emergency lighting to have battery powered emergency lighting installed.

Luminaires have to be manufactured to BS 4533: Part 102.22: 1990 this being the equivalent of the harmonized European standard EN 60–598.2.22: 1989; a typical lighting fitting is shown in Fig. 15.2.

Fig. 15.2 Typical surface mounting emergency lighting luminaire. (Courtesy of Choride Bardic Ltd.)

Types of supply

The types of emergency lighting provided comprise three categories:

- Nonmaintained.
- Maintained.
- Sustained.

In the nonmaintained lighting system (given the category NM) the emergency lamps are not normally energized while the mains supply to the normal lighting is still available. Self-contained luminaires are available which contain their own battery, control gear, battery charger and lamp. The lighting fitting is equipped with a sensing device to switch the lamp on when the normal supply fails. The battery charger recharges the battery on restoration of the power supply (Fig. 15.3).

In the maintained lighting system (given the category M) the emergency luminaire is energized continuously from the mains supply. In the event of a mains failure the supply is automatically switched over to the battery supply. The lighting fittings can be switched providing the light will come on automatically in the event of a power failure. Where the supply is from a central battery unit a separate distribution system will be required

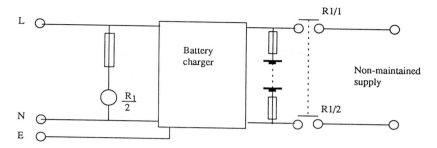

Fig. 15.3 Basic nonmaintained supply.

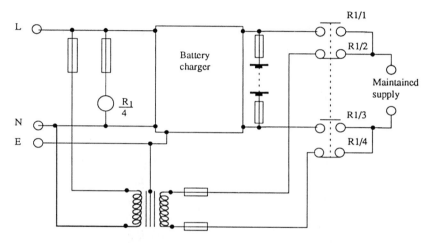

Fig. 15.4 Basic maintained supply.

so that only the emergency lighting fittings are on with a power failure. Using maintained lighting ensures that with a mains failure the level of illumination in the emergency mode remains the same as in the normal mode (Fig. 15.4).

The central emergency unit can be an engine driven generator to give the power for emergency lighting, however, where it is the sole source of supply it should run up to full output in 5 s. It may be possible to increase this time to 15 s where the people in the building are familiar with the building and the escape routes.

Where a large amount of emergency power is required and light is required immediately, as is the case with a hospital theatre, a combination of central generating plant and local battery supplies is used, the latter reacting instantaneously to a power failure.

It is also possible to obtain luminaires similar to those used for non-maintained lighting, but with the additional feature that the mains supply to the lighting fitting can be switched. In the event of a mains failure the emergency supply is automatically switched on to the lamp.

In the sustained system the luminaire is equipped with two lamps, one of which is supplied by the mains and the other supplied from the emergency supply. The emergency lamp remains in the off position until there is a mains failure, at this point the mains operated lamp goes out and the emergency light comes on. The sustained luminaire can be equipped with a separate battery, battery charger and control circuit, which actuates the emergency lamp in the event of a power failure.

Power supplies

The usual source of power for emergency lighting is batteries, these should not, however, be the type used in motor vehicles. Perhaps the best type of battery is the vented nickel cadmium type, it has a life expectancy of 25 years, is resistant to abuse and temperature variations and has a low maintenance cost, see Figs 15.5 and 15.6. Unfortunately its initial cost is high and it requires a ventilated battery room.

Fig. 15.5 Central battery system control panel showing transformers at each end. (Courtesy of Chloride Bardic Ltd.)

Fig. 15.6 Internal view of a central battery system showing control circuits. (Courtesy of Chloride Bardic Ltd.)

A simple calculation can be made to determine the amount of ventilation required as follows:

$$n_a = \frac{C_b \times I_c}{22 \times R_v}$$

where n_a is the number of air changes required per hour, C_b is the number of cells in the battery, I_c is the charge rate in amperes and R_v is the volume of free air in the room in cubic metres, i.e., deduct the volume of cupboards etc. If the room ventilation is worked out for boost charge conditions it will automatically cover normal working conditions, since the gas given off during a trickle charge is only a fraction of that during a boost charge. The battery capacity required will be dependent upon whether the emergency lighting duration is to be 1, 2 or 3 hours. The number of hours the lighting can be sustained is then added to the system category to give a classification, thus a nonmaintained system having a duration of one hour would then be categorized as NM/1.

Recommendations as to the categories that should be applied to certain premises are given in BS 5266: Part 1: 1988.

Emergency signs

Directional signs have to be provided where the emergency exit cannot be seen and there could be some doubt as to the direction to take to get to the emergency exit. Additionally signs are required at exit from the building. All exit signs and directional signs (unless they are internally illuminated) require illuminating to at least 5 lx and the height of signs should be between 2 m and 2.5 m from the floor.

BS 5499: Part 1: 1990 calls for all newly-installed signs to have a pictogram together with the appropriate wording as illustrated in Fig. 15.7, the European pictogram signs rely solely on the graphic symbols and exclude any text on the sign. The object being for the purpose and function of the sign to be obvious to all nationalities (see Fig. 15.7). BS 5499: Part 3 specifies the construction requirements for internally illuminated signs, specific requirements for cinemas and theatres are given separately.

Wiring

Since emergency lighting is provided to illuminate escape routes during a fire it is essential that the cables used to supply the emergency lighting fittings are either resistant to fire or are protected so that they have a fire resistance equivalent to the rating of the system for example, one-hour, two-hour or three-hour fire resistance.

The wiring to a self-contained luminaire can be normal wiring, such as PVC, since the luminaire contains its own emergency battery supply and will automatically changeover in the event of a mains failure.

Fig. 15.7 Illuminated pictogram emergency exit directional sign. (Courtesy of Chloride Bardic Ltd.)

Mineral-insulated copper sheathed cables manufactured to BS 6207: Part 1 can be considered fire resistant. The degree of fire resistance being dependent upon the type of termination used for the cable. Category B cables manufactured to BS 6387 can also be considered to be fire resistant, an example of this type of cable is the Fire Tuf OHLS cable.

Other types of cable will need additional protection against fire for example, PVC cable manufactured to BS 6004 and PVCSWA & PVC cables manufactured to BS 6346 or BS 5467. Additional protection could be formed by the cables being buried within the building structure, such as buried in the plaster.

It may be possible to install the cables in areas where there is very little risk of fire, but ultimately the cables will connect to the luminaires in the escape routes where the risk of fire is a possibility.

Cable sizes

The sizing of cables must take into account:

- The rating and type of protective device.
- The design current for the circuit.
- Cable grouping.
- Ambient temperature.
- Contact with thermal insulation.
- Voltage drop.
- Thermal constraints imposed by prospective short-circuit current.
- Thermal constraints imposed on the protective conductor.
- Constraints imposed by protection against indirect contact.

As far as voltage drop is concerned, the IEE Wiring Regulations call for the voltage at the terminals of the equipment to be not less than the maximum specified in the British Standard for the equipment. In this respect BS 5266: Part 1 specifies that the maximum voltage drop allowed be 10% of the system nominal voltage with the maximum current flowing in the conductors at the highest operating temperature of the conductors. Calculations of the other items in the list have already been given in other chapters of the book and it is therefore unnecessary to repeat them here.

Segregation

The cabling for emergency lighting circuits constitutes a category 3 circuit and as such to comply with the IEE Wiring Regulations it must be segregated from category 1 and 2 circuits. BS 5266: Part 1 specifies that for escape lighting the minimum distance between emergency lighting circuits and circuits of other services must be at least 300 mm. It then specifies that where this separation is not possible MICC cables should be used, these being sized to the exposed-to-touch current rating tables in the IEE Wiring Regulations. BS 6387: 1991 gives the specification for performance requirements for cables required to maintain circuit integrity under fire conditions and any cable having a BASEC certificate of assessment complying with this standard would be just as acceptable as MICC cable.

Installation materials

If conduit, trunking, ducting, channel or junction boxes are to be used whether or not they are metallic they should be of adequate strength and resistance to fire. Details of the specifications that these materials should comply with will again be found in BS 5266: Part 1. Any ducting or trunking etc. used for emergency lighting circuits should be marked to indicate its use.

Inspection and tests

The periods between inspection and tests recommended in BS 5266: Part 1 are:

- Daily.
- Monthly.
- Six-monthly.
- Three-yearly.
- Subsequent annual test.

The inspection to be carried out and tests to be made are fully detailed in BS 5266, which also contains sample completion and inspection and test sheets, a log book to be retained on site will also be required. The tests outlined in BS 5266: Part 1 are in addition to any required by the IEE Wiring Regulations.

Product registration

The Industry Committee for Emergency Lighting (ICEL) offers a scheme of product registration for the purpose of providing assurance to the user that the products registered under the scheme have previously been:

certified to the appropriate national and international standards, that the manufacture of the product is carried out in a facility operating a recognized scheme of quality assurance (for instance BS 5750) and that performance claims made for the product are valid and representative of typical production of the product.

Products registered under the new ICEL scheme may be marked with the ICEL product registration mark.

Chapter 16
Earthing

The system neutral is earthed for two reasons; the first for system operation and protection and the second to reduce the risks to life.

There are four ways of dealing with the neutral at the transformer:

(1) Isolating the neutral.
(2) Earthing the neutral through a resistance (resistance earthing).
(3) Earthing the neutral through a coil (arc suppression earthing).
(4) Solidly earthing the neutral.

Items listed under (1)−(3) would all form the source of an IT system, whereas item (4) would form the source of a TN or TT system. (These systems are defined later in the chapter.)

If the neutral is isolated from earth no fault current can flow from the system and the faulty phase then takes earth voltage. This causes a rise in voltage on the other two phases of approximately $\sqrt{3}$ phase voltage and an imbalance in the capacitance currents from the three phases results in a current flow from the two healthy phases through their insulation to earth, the current returning to the transformer by way of the fault. The result can be intermittent arcing which can impose additional electrical stresses on the insulation and cause further faults to occur.

Resistance earthing usually subjects the healthy part of the system to a voltage increase again of approximately $\sqrt{3}$ phase voltage during a single phase earth fault. This usually only persists for a short time during which the fault current is restricted to a reasonable value by the neutral resistance.

The arc suppression coil method of earthing, more commonly known as the Peterson coil, is a method used to gain the advantage of the isolated neutral without the problem of unbalanced currents producing intermittent arcing. This method is illustrated in Fig. 16.1.

A fault to ground on one phase causes a single phase voltage to appear across the coil resulting in a current (I_L), which lags the line to neutral voltage by approximately 90°, and two capacitance currents i_1 and i_2 from the two sound phases each of which are equal to about three times the normal phase to earth capacitance current. This capacitance current leads the phase to neutral voltage by about 90°. Since these currents also pass through the fault the resulting current I_f is small and almost in phase with the phase-to-neutral voltage, as illustrated in Fig. 16.2.

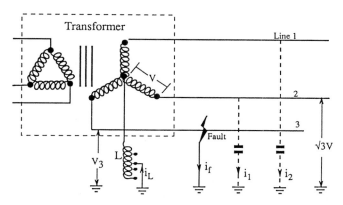

Fig. 16.1 Arc suppression method of earthing neutral.

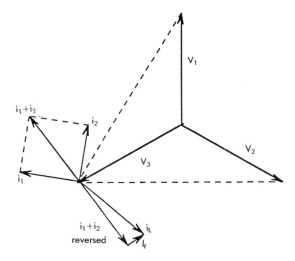

Fig. 16.2 Vector diagram for condition shown in Fig. 16.1.

If the reactance in the neutral is adjusted (tuned) so that it resonates with the system capacitance, in theory, the fault current should be zero; in practice the fault current can never be zero because of the small energy current taken by the coil and some resistance at the fault, but this does not impair its effectiveness as an arc suppressor.

When a Peterson coil or neutral resistance is used and a fault occurs the voltage on the healthy lines rises to approximately $\sqrt{3}$ times the normal phase to neutral voltage.

In the UK the method adopted for protection against earth fault currents is solid earthing, that is, by providing a low resistance circuit between the supply source earth and the exposed conductive parts of final circuits. The circuit provided must be of sufficiently low impedance to enable a high enough current to flow to operate the protective device within times stipulated to comply with the IEE Wiring Regulations. The actual disconnection time required for each circuit is dependent upon how the circuit is to be used, the voltage to earth and the environmental conditions.

Within the equipotential zone and for normal environmental conditions, where the voltage to earth is between 220 V and 277 V, the disconnection time for socket-outlet circuits or circuits supplying hand-held equipment is 0.4 s. The disconnection time for fixed equipment being 5 s. Where socket-outlet circuits are installed with the protective conductor complying with Table 41C the disconnection time allowed is also 5 s. Disconnection times can vary for different environmental conditions or for where the equipment is not within the equipotential zone. Circuits supplied at 110 V between phases, which gives 55 V to earth for a single-phase supply or 63.5 V for a three-phase supply, can have a disconnection time of 5 s for either socket-outlet circuits or fixed equipment.

In general, the above requirements mean that a protective conductor has to be installed to each piece of equipment, except that, where the whole installation or circuit consists entirely of class II equipment and the circuit is under effective supervision so that no changes are made that would degrade the effectiveness of the class II insulation, a protective conductor need not be installed to a fixed equipment circuit. As not all equipment is double insulated there is always the possibility that the unwary may use socket outlets for class I equipment unless further precautions are taken, such as the installation of a completely different type of outlet. Class I equipment is that provided with basic insulation only and provision for the connection of protective conductor.

Types of earthing system

There are five types of earthing system in use, listed in the IEE Wiring Regulations as described below.

TN-C system

Here one conductor only is used for both the neutral and the protective part of the circuit. To provide the facility for bonding exposed and extraneous conductive parts, the appropriate cable is the protective earth neutral (PEN) type, in which the insulated phase conductors are surrounded by a conductive sheath providing the neutral and protective conductor element. It should be noted that, with this system, residual current devices are prohibited as they will not detect an earth fault (Fig. 16.3).

TN-S system

In this system neutral and protective conductors are completely separate. The latter may take the form of a conventional conductor or be provided by the continuity of trunking, conduit, the metal sheath or armour of a cable in the consumer's installation and, on the supply side, by a separate conductor on overhead systems or cable armouring on underground distribution. The TN-S system is the most commonly used one in the UK although, due to the variations in the specific impedance of the earth's surface and, therefore, the difficulties in obtaining a good earth at substations, it is being superseded by the TN-C-S system (Figs 16.4 and 16.5).

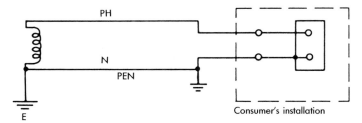

Fig. 16.3 TN-C system.

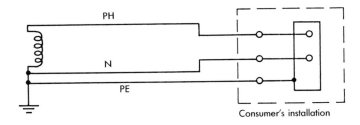

Fig. 16.4 TN-S system.

TN-C-S system

This should not be referred to as the PME (protective multiple earthing) system, since PME is a type of supply. This system is one in which the neutral and protective conductors are combined in the supply side part of the system with separate conductors in the consumer's installation. Satisfactory earthing is obtained, as the name implies, by the use of a number of earth plates or electrodes connected directly to the supply system neutral (Fig. 16.5).

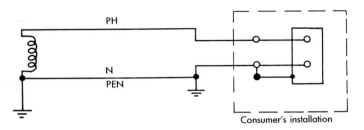

Fig. 16.5 TN-C-S system.

TT system

In the TT system both the regional electricity company and the consumer must provide an earth electrode at appropriate locations, the two being completely separate electrically. The consumer's protective circuitry is connected to his earth plate, and protection may be provided by means of a residual current device (the method preferred by the IEE Wiring Regulations) or by overcurrent protection. Fault voltage operated devices were a permissible means of protection in TT systems until January 1986 (Fig. 16.6).

The above terms are taken from the 16th Edition of the IEE Wiring Regulations Part 2 – Definitions, where the alphabetical coding is explained in detail.

From the above it is clear that, before the design of an installation is

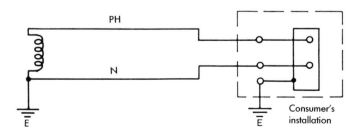

Fig. 16.6 TT system.

commenced, the type of system must be ascertained as it may affect the requirements for protective devices.

IT system

Unlike the previous systems, the IT system, which is not permitted on the low voltage system by the Electricity Supply Regulations 1988 (as amended), does not rely upon earthing for safety as the supply side is either completely isolated from earth, or is earthed through a high impedance, and only the exposed conductive parts of the installation connected to an earth electrode. Consequently, should an earth fault occur, it is impossible for a fault current to flow back to the source of supply, or alternatively it is limited by the impedance and, therefore, dangerous voltages should not develop on an installation's exposed conductive parts (Fig. 16.7).

Main earthing connection

The responsibility for earthing an electrical installation rests with the consumer. Where the electricity supply is provided at low voltage the regional electricity company will provide an earthing terminal, the cost of which is usually included in the cost of providing the supply. Where the electricity supply is provided at high voltage to the consumer then the earthing of the star point of the consumer's transformer is the responsibility of the consumer.

Every installation has to be provided with a main earthing terminal to enable: circuit protective conductors, the main equipotential bonding

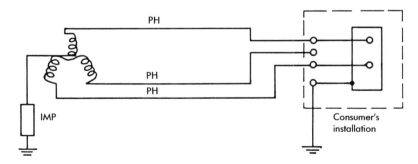

Fig. 16.7 IT system. The Electricity Supply Regulations 1988 (as amended) do not permit this type of supply to be used on the low voltage network distributed to consumers.

conductors and any functional earthing conductors or lightning conductors to be connected to the means of earthing by a main earthing conductor.

The size of the main earthing conductor is determined by using either Table 54G or the formula given in Chapter 54 of the Regulations:

$$S = \frac{\sqrt{I_f^2 t}}{k}$$

where S is the minimum cross-section area required for the earthing conductor, I_f is the prospective earth fault current at the installation's main switch determined from the formula U_0/Z_S, t is the time taken for the protective device at the origin to disconnect the fault current I_f and k is a factor dependent upon the conductor's material and insulation.

Where the earthing conductor is buried in the ground it is subject to a minimum size depending upon whether or not it is protected against mechanical damage and/or corrosion (Table 54A in the IEE Wiring Regulations). Aluminium or copper-clad aluminium conductors are not allowed to be used for the underground connection to an earth electrode.

The means of earthing is dependent upon the type of system of which the installation will be a part. For an installation which is part of a TN-S system the means of earthing is the metallic sheath and/or armouring of the supply cable. For an installation which is part of a TN-C or TN-C-S system it is the PEN conductor of the supply cable which is the means of earthing. For an installation which is part of a TT or IT system the means of earthing is an earth electrode at the installation.

Where the supply to the installation is low voltage and the installation is part of a TN-C or TN-C-S system, the responsibility for making the earthing connection rests with the electricity supplier.

Circuit protective conductors

The well-known term ECC (earth continuity conductor) has now been replaced by 'circuit protective conductor' (CPC), and is applicable to any conductor or conductive material installed for the return path of an earth fault current should a fault occur on a circuit or on equipment connected to that circuit. The CPC may consist of a conventional conductor or be provided by the conduit, trunking, tray, etc., installed for the purpose of supporting the live cables, provided that the latter has an equivalent resistance to the calculated requirement for a conductor. As it is unlikely, in the case of conductors, that a calculated size will match a standard rating, the next larger cross-sectional area should be installed while, due to possible deterioration of joints and other causes, when supporting equipment is utilized for the CPC it is strongly advised that the resistance should be well below the maximum or a separate conductor installed.

Main equipotential bonding

Although the term 'main equipotential bonding' is not included in Part 2 Definitions of the Regulations, reference to Regulation 413−02−02 indicates clearly the application and purpose of this requirement, i.e., to bond together and connect to earth conductive parts that are not directly associated with an electrical installation but which could provide a circuit for a fault current under certain circumstances.

To comply with the IEE Wiring Regulations, the cross-sectional area of this conductor must not be less than half that of the main earthing conductor, subject to a minimum of 6 mm^2 and need not exceed a maximum of 25 mm^2 for a copper conductor unless the supply is PME. Where the supply is PME the equipotential bonding conductors are sized according to the equivalent size of the supply's neutral earth conductor, as given in Regulation 9 of the Electricity Supply Regulations 1988 (amended) or Table 54H of the IEE Wiring Regulations. It should be specially noted that this bonding is not for the purpose of providing an earth connection, but for bonding other services within the consumer's property to the main earthing connection. It is also essential to install it in such a manner that, should one service be disconnected, the bonding of the remainder will not be affected; this may be achieved by the installation of one continuous conductor and looping into each service connection.

Reference should be made to Regulation 547−02.

Supplementary bonding conductors

The main objective of all bonding is to create zones within which the various conductive parts, whether exposed or extraneous, remain at essentially the same potential under fault conditions. As many items manufactured from conductive material may be installed throughout a building remote from the origin of the installation, provision is made in the Regulations for this eventuality by reference to the supplementary bonding conductors (Regulations 413−02−27, 28 and 547−03). If the objective is kept in mind, the requirement will often be self-evident, but a fact which may possibly be overlooked is that other facilities are not compelled to install electrically continuous services and, therefore, insulated sections or joints in pipework etc. may have been introduced which eliminate the effectiveness of the main equipotential bonding. In these cases, the responsibility lies with the designer and installer to ensure that supplementary bonding conductors are applied. Apart from special situations, the only time that it is considered necessary for an equipotential bond between an exposed conductive part and an extraneous conductive part, is when the circuit cannot be disconnected within the specified time and it is decided not to use an RCD as protection. In such circumstances the bonding conductor's minimum size is determined by Regulation 547−03; the resistance of the bonding conductor being confined to $50/I_a$ where I_a is the current needed to disconnect the circuit within five seconds as called for by Regulations 413−02−27 and 28.

The different conductors are indicated in Fig. 16.8.

Automatic disconnection devices

Main circuit protection devices, either fuses or circuit-breakers, can be used to clear earth fault currents provided that the maximum impedance (Z_s) related to the device is not exceeded, but this entails, even though for short periods of time, relatively high fault currents and protective conductors capable of handling these without dangerous increases in the temperature of the conductors. However, conditions may be such that alternative methods of earth fault protection are necessary, e.g. where earthing is poor, giving rise to a high external impedance (Z_E), or on extensive installations, when it may be difficult to comply with the maximum Z_s requirements.

Regarding Z_E, regional electricity companies tend to quote a range, between $0.35\,\Omega$ and $0.8\,\Omega$ and, if the higher value is taken, for a 5 s period of disconnection, BS 88 fuses above 63 A, BS 3036 fuses above 60 A and MCBs or MCCBs above 30 A (although this depends upon the type) are totally inadequate for overall protection by themselves. Even at the lower ($0.35\,\Omega$) level of the range quoted, HBC fuses of the type referred to, above 125 A, would not clear a fault current within the prescribed time limit. Obviously, any impedance in the installation's live and protective conductors adds to Z_E (to obtain Z_s) and, therefore, on large installations with numerous submains and distribution centres in series, Z_s may not be achievable within the maximum limits, and each area must be dealt with separately.

Where prohibiting circumstances arise, it is necessary to install devices which can detect and, if necessary, disconnect a faulty circuit before dangerous situations are created.

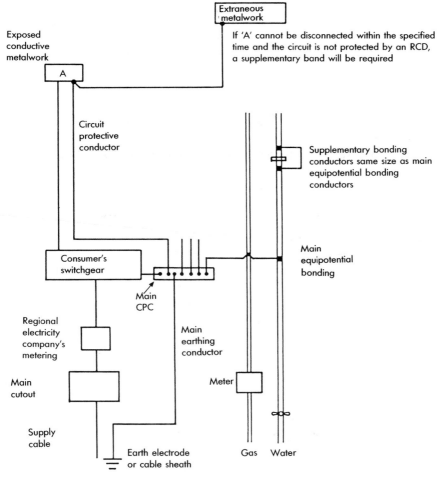

Fig. 16.8 Earthing and bonding requirements.

Similar provision may also have to be made in industries which include dangerous processes (chemical or explosive) or in which greater danger might be created by an unexpected shut-down.

Residual current devices

A residual current protective device is defined in the IEE Wiring Regulations as a mechanical switching device or association of such devices intended to cause the opening of contacts when a residual current attains a given value under specific conditions – the definition does not insist that the main circuit shall be disconnected directly by the device(s) nor that disconnection shall be effected at all, subject to certain conditions, e.g. that safety must be considered at all times. For the majority of small to medium-sized installations, whether domestic, commercial or industrial, what was known as the earth leakage circuit-breaker but is now referred to by many manufacturers as the residual current circuit-breaker (RCCB) is the best-known and most-used device. It is a direct-acting unit, combining the overcurrent protection function and the detection and disconnection of an earth fault. BS 3871 covers MCBs rated at up to 100 A and manufacturers combine these with residual current tripping facilities of various levels, usually 10, 30 and 100 mA ratings. Consequently, these are suitable not only for all domestic installations but also for many commercial and industrial ones. They also provide the advantage that the tripping period is limited by the above standard to 0.1 s which is well within the limits imposed by the IEE Wiring Regulations.

In a larger installation certain types of electrical equipment have inherent leakage currents, and this must be taken into consideration in choosing a tripping rating or, alternatively, circuits must be designed in such a way that the maximum inherent leakage current is well below the RCCB tripping current.

Due to the ever-increasing size of commercial and industrial installations, the combined units referred to are not capable of handling the power supplies demanded at the origin of supply and, therefore, alternative types of equipment such as switch-fuses, MCCBs or standard types of circuit-breaker are required. As the first is not a totally automatic disconnector, a residual current device (RCD) can only be used to indicate an earth fault and the operation of the fuses will depend upon the time/current characteristics in exactly the same way as with an overcurrent. In this case, the installation of a suitably rated RCD (low mA rating) is used to give audible or visual indication of an earth fault. It is more usual, however, to apply residual current protection to final circuits.

For the higher power requirements, circuit-breakers of the types mentioned are preferable and, in these cases, assemblies of current transformers and relays are relatively easily incorporated in switchgear to provide both overcurrent and earth fault protection. This has the advantage that both current and tripping time levels are adjustable within limits to suit the requirements of the user while the standard BS 3871 equipment has no such facilities.

Residual current protection/detection devices rely on the principle that all of the current flowing through the detection coil into the load returns through the detection coil. Since the currents are in opposite directions the flux they generate in the detection coil cancels out so that no voltage is generated across the tripping mechanism and the unit does not trip. When a fault occurs some, if not all, the current flowing into the fault returns to the source via the protective conductor or an earth path. This causes an imbalance between the current flowing into the fault and that returning via the protective conductor or earth (since these do not pass through the detection coil), this imbalance triggers a tripping mechanism.

The residual current device (RCD) does not limit the magnitude of the fault current at the instant of an earth fault, the fault current is only limited by the system impedance Z_s. Additionally, the RCD does not limit overload current or short-circuit current since the current still returns through the detection coil and thus an imbalance is not created.

It is important to ensure that protective conductors or earthing conductors are not passed through the detection coil in the RCD since this would stop the unit tripping. It is also essential to ensure that the following conductors pass through the detection coil of the RCD:

- Both the phase and neutral for single-phase supplies.
- All three phases of three-phase supplies.
- All three phases and neutral of three-phase and neutral supplies.

If the neutral of three-phase and neutral supplies is not passed through the detection coil nuisance tripping will occur every time a single-phase circuit is switched on.

The only way discrimination between RCDs which are in series in a distribution system can be achieved, is by using different time delays in each device. The time delay, delays an RCD tripping giving the RCD downstream chance to disconnect the faulty circuit first. However, the time delay stops the RCD from giving protection against electric shock and should not therefore be used in a final circuit.

In general, it is better to limit the use of RCDs to final circuits and not install them at the mains position. This, however, is not always possible when the installation is part of a TT supply.

An RCD provides inherent selectivity in that it protects only the circuit in which it is installed and, consequently, the designer/installer is able to arrange circuitry and connected equipment so that nuisance tripping is avoided (Fig. 16.9).

Fault voltage operated protective device

This was the first type of earth leakage circuit-breaker introduced into the UK from Europe and has been available for approximately 50 years. The

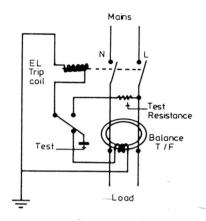

Fig. 16.9 Circuit diagram for current-operated RCCB.

main application of this type is for protection against indirect contact, as the operation is effected by the detection of a voltage increase on the metalwork of an installation. Provided that the earth electrode for the unit is outside the effective resistance area of other earth electrodes, fortuitous earth paths occurring elsewhere on an installation will not defeat the operation of the trip-coil; they may, however, lead to high fault currents, with the attendant danger of fire and, also, unless metalwork associated with other circuits similarly protected is completely insulated, all units in an installation will operate on a rise of potential. Consequently, fault voltage operated types may not protect from fire risks and, practically, are only suitable for small installations such as domestic ones.

Due to such considerations as those above, the IEE Wiring Regulations now exclude their use. Figure 16.10 shows a typical arrangement.

Earth proving systems

In many commercial and industrial installations, metal conduit, trunking and tray are used to provide all or part of the protective circuitry, and the integrity of the continuity must be maintained to comply with the regulations, whatever type of protection is installed. However, building vibration may loosen connections or corrosion take place at joints, and deterioration may not be detected even with frequent inspections. To avoid such situations arising, particularly in known hazardous processes, earth proving systems are employed which constantly monitor the effec-

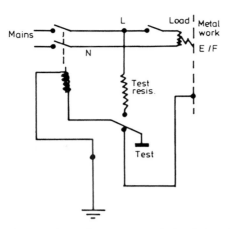

Fig. 16.10 Circuit diagram for voltage-operated circuit-breaker protection. This type is no longer acceptable to conform with the requirements of the IEE Wiring Regulations.

tiveness of protective conductors or circuitry; the detection of break-down in the protective circuit is then used either to disconnect the associated supply or to give warning of the failure so that appropriate action may be taken to rectify matters as soon as possible.

One method adopted for this is the installation of a small step-down transformer with screened windings, giving an extra-low voltage output, across the supply; the secondary is then connected to each end of the protective conductor through a relay in the monitoring unit. If the resistance of the protective conductor increases through corrosion, breakage, etc., the relay is de-energized and the changeover of the contacts disconnects the main supply or gives some form of visual or audible indication. Electronic equipment is also available which serves the same purpose but which, due to the relatively higher cost, tends to be more used in flameproof or similar high-risk situations, especially where intrinsically safe circuits and equipment are installed, rather than on general commercial and industrial installations.

Examples checking protective conductor sizes

The values used in the following examples have been taken from the information supplied in Chapter 12 and the Handbook on the IEE Wiring Regulations.

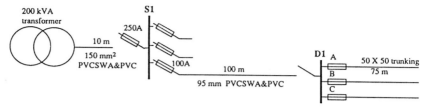

Fig. 16.11 Factory distribution.

Example 1

Figure 16.11 shows a factory distribution scheme which uses HRC fuses. Determine that the armour of the 95 mm² cable back to S1 is suitable for the thermal stress imposed by a fault to earth on the busbars of distribution board D1, if the armour area of the 95 mm² cable is 160 mm² and the k factor for the steel wire armour is 51.

Working

Again the worst case will occur if the conductors of the circuit reach the average of the permitted operating temperature and limit temperature for the conductor's insulation, i.e., 115°C.

From Chapter 12

The impedance of the transformer is $0.014 + j0.0423$ which results in an impedance of $0.0446\,\Omega$.

The earth loop impedance of the 150 mm² cable is $1.04\,\Omega$ per 1000 m. The earth loop impedance of the 95 mm² cable is $1.52\,\Omega$ per 1000 m.

The total earth loop impedance from the transformer up to distribution board D1 equals $0.0446 + (1.04 \div 100) + (1.52 \div 10) = 0.207\,\Omega$.

The fault current at D1 will be $240\,\text{V}/0.207 = 1159\,\text{A}$

It is now necessary to determine how low it will take the 100 A fuse to disconnect the circuit. This is done by looking at the characteristic of the 100 A fuse, finding 1159 A on the current axis drawing a vertical line up to the characteristic then horizontally to the time axis. This gives a disconnection time of 0.175 s.

This information is now substituted into the formula

$$S = \frac{\sqrt{I_f^2 t}}{k} = \frac{\sqrt{1159^2 \times 0.175}}{51} = 9.5\,\text{mm}^2$$

which is much less than the cable armour area of $160\,mm^2$, therefore the armour is satisfactory and would still be satisfactory even if the fuse rating was increased to $200\,A$.

Example 2

If a $6\,kW$ infrared heating unit is to be installed from distribution board D1 over the steps into the building and is to be wired in 2.5/1.5 twin and CPC cable, will the CPC be satisfactory if the cable length is $50\,m$ and the protective device size is $25\,A$ HRC? The k factor for copper CPC enclosed with live conductors is 115.

Working

The phase earth loop impedance up to the busbars of D1 was determined in example 1, which was $0.207\,\Omega$, this is the Z_E for any circuit being taken from D1.

Note that Table 54G cannot be used since the CPC is not the same size as the phase conductor.

From Chapter 12; R per $1000\,m$ for $2.5\,mm^2$ cable $= 10.226\,\Omega$ and $1.5\,mm^2$ cable $= 16.698\,\Omega$.

Z_{inst} for the circuit will be $\dfrac{50\,(10.226 + 16.698)}{1000} = 1.346\,\Omega$

Using the formula from Chapter 12, $Z_s = Z_E + Z_{inst}$

$Z_s = 0.207 + 1.346 = 1.55\,\Omega$

$I_f = \dfrac{240\,V}{1.55} = 154.8\,A$

Now determine how long it will take the $25\,A$ HRC fuse to disconnect the circuit. From the fuse characteristic $t = 0.39\,s$.

$S = \dfrac{\sqrt{I_f^2 t}}{k} = \dfrac{\sqrt{155^2 \times 0.39}}{115} = 0.84\,mm^2$

The cable CPC is satisfactory.

It should be noted that the first example was for a three-phase circuit and the second for a single-phase circuit, but both calculations were carried out the same way.

It should also be noted that it is easier to satisfy the IEE Wiring Regulations for protection against indirect contact and protective conductor sizes when the supply is from the factory transformer rather than from the regional electricity company's low voltage network.

Chapter 17
Lightning Protection and Surge Suppression

The frequency, duration and intensity of lightning discharges is less in the UK than in many tropical countries, but there are very few locations in the UK which are immune from the effects of a lightning discharge. Again in the UK there are more discharges in the east than in the west and more occur in the south than in the north.

Damage caused by a lightning flash is by the 'return stroke' which is that part of the flash during which charged cells in a thunder cloud discharge themselves to earth. Currents in such a discharge vary from about 2000 A to 200 000 A, but only about 1% of these lightning strokes exceeds 200 000 A and approximately 50% exceed 28 000 A in the UK.

As the discharge seeks the lowest impedance a building, particularly one containing large amounts of metal in the structure, is susceptible to lightning strikes and the electrical installation which of course, is designed to be of a low impedance, is even more prone to a lightning strike as a result of a side flashover.

If a protection system is installed which effectively envelopes the building so as to provide a highly conductive path to earth, and it is adequately bonded to the electrical installation so as to minimize flashover, it will protect both the structure and the electrical installation.

Contrary to popular belief the thermal effect of a discharge on the lightning conductor's protection system is usually negligible as this is limited to the temperature rise on the lightning conductors through which the current passes and as the cross-sectional area of the conductor itself is normally designed to prevent mechanical effects, it means that in practice the rise in temperature is limited to approximately 1°C. This figure assumes that the conductor material is copper, a small increase on this figure of 1°C occurs if other less conductive materials are used.

It is important to provide sufficient mechanical strength for the lightning conductor protection system, since considerable mechanical forces are produced by the high current discharging along parallel conductors or through sharp bends. All components of the system should therefore be of robust construction and securely fastened to the structure.

Shock waves occur close to the lightning stroke caused by the explosive expansion of air, the temperature of which rises sharply due to the lightning stroke, and this can for instance, readily dislodge roof tiles. In a similar manner side flashes within the building resulting from insufficient or inefficient bonding can cause damage.

It must also be remembered that as a lightning discharge raises the potential of the system above that of earth it produces a high voltage gradient between the system's earth electrode and the surrounding soil. It is absolutely essential therefore that the system must have a substantial earthing arrangement of low ohmic value to counter this electrical effect.

The protective system's main function is to divert to itself a lightning discharge which may otherwise damage the structure, and it is generally acknowledged that the range over which a system attracts a lightning flash is a function of the severity of the charge, and that vertical and horizontal air terminations have an equivalent attraction to the charge.

Need for protection

In many cases the requirement for a protection system is self-evident, but there are cases where it is not always easy to make a decision. Reference should therefore be made to British Standard BS 6651:1992. The need for protection can be established by consulting subsection 9 of Section 2 of the standard. Tables 6−12 in the standard detail the various weighing factors to be considered and Table 13 is an example of the calculations necessary to evaluate the need for protection.

The criteria for calculation are:

- Relationship between thunderstorm days per year and lighting flashes to ground per km^2 per year (Ng).
- Comparative probability of death for an individual per year of exposure.
- The use to which a structure is put.
- The type of building construction, i.e., brick or steel reinforced concrete.
- The contents of the structure, i.e., contents are inflammable.
- The degree of isolation of the structure, i.e., in the city centre or in open country.
- The type of surrounding country, i.e., flat or mountainous.
- The height of the structure.

To determine the probable number of lighting strikes that may be made to a structure each year BS 6651 gives the following formula:

$$P = A_c \times N_g \times 10^{-6}$$

where P is the probable number of strikes, A_c is the collection area in m^2, and N_g is the number of flashes to ground per km^2 per year. The collection area A_c is determined by the plan of the structure extended in all directions by its height, this is illustrated in Fig. 17.1 and it can be seen that the collection area is:

$$LW + 2LH + 2WH + 4\left(\frac{\pi\, H^2}{4}\right)$$

The last part of the expression picks up the area of the four corners.

Having determined the probable number of lightning strokes, it is then necessary to apply weighing factors obtained from Tables 8(A) to 12(E) from BS 6651: 1992. These weighing factors are multiplied together and then multiplied by the probability factor previously worked out, for example the formula becomes:

Overall risk factor = P × A × B × C × D × E

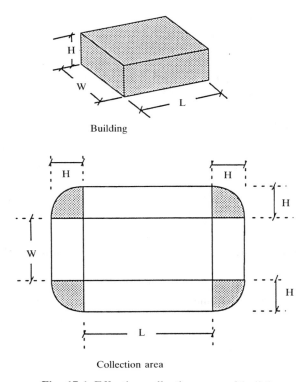

Building

Collection area

Fig. 17.1 Effective collection area of building.

The answer is then expressed as a reciprocal, so the ultimate formula becomes:

$$\text{Overall risk factor} = \frac{1}{P \times A \times B \times C \times D \times E}$$

This gives a ratio for instance, 1 in 200 or 1 in 300 000.

Having applied the formula and established the ratio it is necessary to interpolate the result. Should the result exceed 1 in 100 000 it would be considered unnecessary to provide protection to the building. However, some anomalies still arise for instance, an unusually tall structure of nonconductive material for example, masonry, should be protected by virtue of its height, even if the risk factor is less than 1 in 100 000. Structures containing explosive substances are, of course, subject to additional considerations and fall into a separate method of calculation.

Zone of protection

The zone of protection afforded to a structure by a protective system fitted to it is calculated by three different methods:

- Structures up to 20 m high.
- Structures higher than 20 m.
- Vulnerable structures such as, those containing explosives.

Structures up to 20 m high

For structures up to 20 m in height, the zone is taken as a cone with its apex at the top of the lightning conductor attached to the structure and its base on the ground. Most buildings are of a greater area than can be protected by a single vertical lightning conductor, making it necessary to install several such conductors connected together by a horizontal conductor, say along the ridge of the building. In the latter case the zone of protection is taken as the volume of a cone with its apex at the horizontal conductor, moving from end to end.

The term zone of protection is defined in BS 6651 as the 'Volume within which a lightning conductor gives protection against a direct lightning strike by diverting that strike to itself'. The protective angle being between the side of the cone and the vertical to its apex (See Figure 17.2). The size of the angle cannot be accurately stated as it depends upon the severity of the stroke and the presence of other conductive elements providing independent paths to earth. It must be assumed

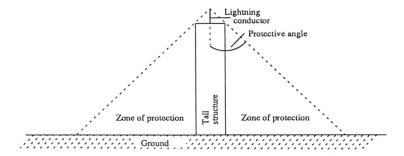

Fig. 17.2 Zone of protection of a tall structure.

that the protection afforded by a conductor increases as the protective angle decreases.

Structures higher than 20 m

Due to the variation in the protective angle mentioned above, where a building with a height in excess of 20 m is likely to be struck by lightning on the side of the building, the protective volume is calculated by the 'rolling sphere' method set out in A5 of Appendix A of BS 6651. The method of calculating the number and positions of horizontal conductors is similar to that used for structures less than 20 m high and is fully detailed in Section 18 of the British Standard.

Vulnerable structures

For structures of exceptional vulnerability, i.e., those containing explosive or flammable materials, the design methods are set out in clause 21 of BS 6651. Where possible such structures should be protected by a system of suspended conductors placed above the structure, but using the lower angles of protection given in BS 6651.

Where there is no risk involved in discharging the lightning strike over the surface of the structure a network of roof conductors of some 10 to 15 m apart (or closer if the risk is greater) connected to vertical down conductors may be used. The earth electrodes of each vertical down conductor must be interconnected by a ring earth electrode, and great attention should be paid to bonding all metallic parts in the structure to the lightning conductor system.

Component parts: protective system

The component parts of the lightning conductor protective system are:

- Air terminations.
- Down conductors.
- Metallic parts in contact with the ground.
- Test points.
- Earth electrodes.

Air terminations

Air terminations can comprise vertical or horizontal conductors (or both) fitted to the roof of the structure in such a manner that no portion of the roof is more than 5 m from a horizontal conductor. Care must be taken to avoid sharp re-entrant loops around projections such as parapet walls when connecting the roof terminations to the down conductors as illustrated in Fig. 17.3.

Down conductors

The down conductors are the vertical conductor section of a lightning protection system, fitted to the sides of the structure and spaced round the perimeter of the building at 20 m intervals for buildings not exceeding 20 m in height and at 10 m intervals for buildings higher than 20 m.

The down conductors should be spaced evenly along each side of the building starting at the corners. The stanchions of steel framed buildings or the steel reinforcement in concrete structures should be interconnected with the rest of the structure and may then be used as the lightning protective system down conductors. The design of the system includes the whole structure, any metallic pipes installed should be bonded to the structure top and bottom. Further information is available in Section 15 of BS 6651.

Metallic parts in contact with the ground

All metallic parts in a structure in contact with the general mass of earth should be bonded to the lightning protective system or alternatively

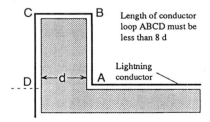

Fig. 17.3 Re-entrant loop example.

completely isolated from the ground. Bonds should be of sufficient electrical and mechanical strength and protected against corrosion. Similarly all joints should have good electrical continuity and be mechanically robust and corrosion resistant.

Test points

Test points should be provided in all down conductors so that the earth terminations can be isolated for test purposes. These must also have the same properties detailed above for joints. The position of the test points is not critical, but they should be in a position where the risk of damage is minimal, limit unauthorized interference and be convenient for testing.

Earth electrodes

Earth electrodes in the form of earth rods or strip electrodes should be connected to each down conductor through the test points. The resistance of each electrode should not exceed $10\,\Omega$ multiplied by the total number of electrodes installed. For example, if 8 down conductors are equally spaced around the building the resistance of each electrode with the test link removed should not exceed $10 \times 8 = 80\,\Omega$. This means that the overall resistance of the earth termination network does not exceed $10\,\Omega$ when the test links are in place and a test is carried out at any point on the system.

The earth electrodes should be disconnected from the lightning protection system, at the test points, whilst the test is made. If the resistance of $10\,\Omega$ cannot be achieved with the designed quantity of electrodes connected to each down conductor additional electrodes can be installed. These can take the form of earth rods or strip electrodes.

Another way of reducing the soil resistivity and hence the earth resistance is to introduce soil-conditioning agents. There are various conditioning agents available, but consideration has to be given to the damage that can be caused to the earthing system by corrosion and the length of time the reduction in earth electrode resistance will be effective. Two agents that can be used with earth rods are Bentonite and Marconite, the latter being the more expensive of the two.

Where earth rods are used care has to be taken that areas do not overlap as illustrated in Fig. 17.4, strip electrodes should preferably be buried at a depth of 1 m, but in any event not less than 0.6 m.

The number and type of each earth electrode should be given on a suitable plate fitted above each test point.

To reduce voltage gradients the earthing conductor from the test point

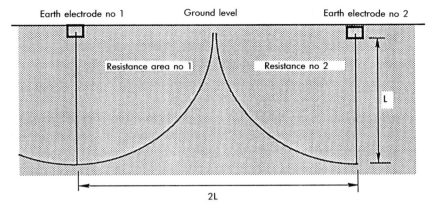

Fig. 17.4 Resistance areas for earth electrodes should not overlap.

to the earth electrode can be insulated with 5 mm thick polyethylene or similar material and the earth electrode and its termination can be installed 1 m below ground level.

One of the biggest problems with lightning strikes is a flash-over from the lightning conductor protective system to other metallic parts at earth potential. This can happen with equipment installed within the structure and to reduce this risk all services within the building must be bonded to the lightning conductor system, this also being a requirement of the IEE Wiring Regulations as illustrated in Fig. 17.5.

Materials and dimensions

The type and specification of materials to be used for the lightning conductor protection system are detailed in Section 1 clause 6 of BS 6651, in particular Table 2 lists all the British Standards for all the various components used in the lightning conductor system including nuts, bolts, washers and screws; typical designs with dimensions are also given for lightning conductor fixings. Characteristics for some of these materials are given in Table 3 (Some materials are illustrated in Fig. 17.6).

Where an installation is required to have a long life, copper or aluminium is recommended with the addition of a protective coating such as, 1 mm thick PVC, although for chimneys a 2 mm thick coating of lead is recommended. Care should be taken, however, when using dissimilar metals in wet or damp situations due to electrolytic action causing corrosion.

Although copper and aluminium are the most widely used materials in the UK, galvanized steel is not prohibited, but the life of the installation

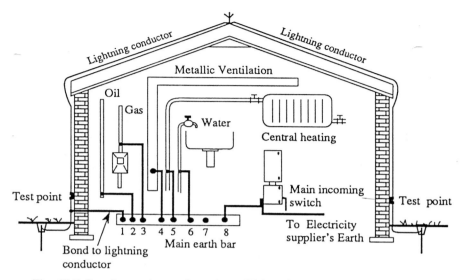

Fig. 17.5 Bonding to internal services. (Taken from *Handbook on the IEE Wiring Regulations* by T.E. Marks.)

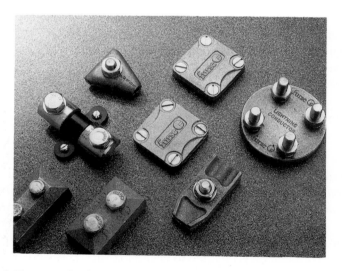

Fig. 17.6 Photograph of materials used for lightning protection. (Courtesy of W.J. Furse & Co Ltd.)

will only be short; this may be acceptable for a temporary installation. As far as the installation from the test point to the earth electrodes is concerned, copper is used for the conductors irrespective of the material used for the lightning conductor system and gunmetal or a similar material is used for clamps and fixings. Where aluminium is used above the test point a suitable bimetal connection should be provided to reduce the effects of electrolytic action between the copper and aluminium conductors. A complete lightening conductor installation is illustrated in Fig. 17.7.

Inspection and testing

An inspection of the lightning protection system should be carried out whilst the installation is in progress to ensure that the materials used and the workmanship employed comply with BS 6551.

The only person who is capable of carrying out the inspection and testing is a competent person who is fully conversant with BS 6651, BS 7430 and any other relevant British Standard or HSE Code of Practice pertinent to the type of structure and the use to which it will ultimately be put and has experience in such installations. Another important quality the inspector must have is a head for heights.

The inspector must have a log book so that a record can be kept of the inspection and tests carried out. Another important item is a sensitive voltmeter so that it can be ascertained that there is no voltage on the lightning conductor system before test points are disconnected.

The inspection and testing should be carried out at intervals not exceeding 12 months, which allows for the time to be varied so that seasonal changes can be taken into account, e.g., long dry spells may affect the resistance of the earthing network.

Fig. 17.7 Photograph of a complete lightning conductor installation. (Courtesy of W.J. Furse & Co Ltd.)

The inspection should include checking the condition of all the conductors, joints and bonds including the earthing conductors where possible. These should be inspected for mechanical strength and corrosion. Any additions or alterations that have been made since the last inspection should be included and a record made of such alterations or additions in the log book.

The testing of the installation will include the continuity of all conductors and the resistance to earth of the earth termination network and each earth electrode. The null balance test can be used for testing the earth electrodes and the earthing network, the continuity of conductors being made with an appropriate instrument. Any large variation between the present readings and previous ones should be investigated to find out the cause.

Records

Records should be kept on site comprising: drawings of the installation indicating in particular the nature of the soil, the position of all of the earth electrodes and test points.

The log book will contain the results of all inspections and tests made and note any alterations or additions to the installation. Additionally it will contain the name of the person or company responsible for the upkeep of the installation.

Maintenance

The tests and inspections carried out will indicate any maintenance that is required, this being detailed in the log book for the installation and any such defects should be rectified quickly. In particular, the earthing of the installation must be maintained at a resistance not exceeding $10\,\Omega$ and any corrosion on joints and clamps shall be dealt with. It is also important to maintain the integrity of all of the fixings since loose ones allow excess movement of the lightning conductor due to the action of the wind, which can cause failure by fatigue of the conductor material.

Maintenance also includes checking whether any items have been added to the building which negate the lightning conductor protection such as, TV or radio masts added to the building after the installation has been completed.

Surge suppression and current chopping

Voltage surges on an electrical installation are frequently caused by external influences such as electrostatic induction created by highly charged atmospheric conditions, direct lightning strokes on the supply system overhead lines, by switching operations on inductive or capacitive equipment on other installations on the same system, and by arcing grounds created by intermittent earths. While the supply authority in the UK normally introduces equipment into its system to counteract the effects of a lightning discharge, anything further is the responsibility of the consumer.

On the size of installation under consideration, it is unlikely that a consumer's normal electrical equipment will be unduly affected (except with regard to MICC cable installations, already referred to) although this may not be the case with computer and similar equipment.

Overhead lines are particularly susceptible to surges created by lightning strokes which, if not diverted safely to earth, often cause serious damage to conductors, insulators and transformers, damage which may extend to associated services. This situation is largely avoided by connecting to the overhead lines some form of arrester or diverter.

For voltages up to 11 kV the most common forms of equipment are arcing horns, aerial/earth wires and multiple-gap arresters. Arcing horns are of simple construction, shaped in the form of an open 'V', the distance between the legs of the horn varying according to the system voltage and corresponding approximately to 1.5 times that voltage. At 11 kV the gap is of the order of 12.5 mm, up to 60 mm at 33 kV. On rural networks the arcing horns are usually mounted across line insulators or the bushings of pole-mounted transformers and connected between the conductors and earth. Due to the short distance between the legs of the horn on the lower voltages, however, they are easily bridged by birds or wind-borne materials and, consequently, it is more usual to install double-gap horns, with the two gaps in series between line and earth (Fig. 17.8).

A more expensive arrangement includes an installation of chokes in the conductors between the horns and the equipment to be protected which, being of low resistance and high reactance, do not affect normal operation but have the effect of reflecting the surges back to the horns. Alternatively, series resistances are placed between conductors and horns to limit the discharge current and prevent complete breakdown of the horns.

The multiple-gap arrester, in one form consisting of a number of concentric cylinders, each one insulated and separated from the adjacent

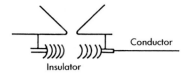

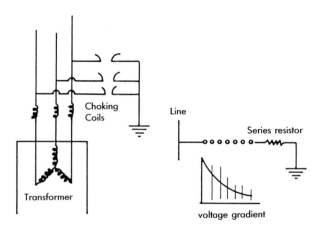

Fig. 17.8 Typical arcing horn/cap lightning arresters.

ones by an air gap, has a greater sensitivity to minor rises in voltage and, again, may have series resistors included.

As will be seen from the above descriptions, gap arresters are really diverters as a true arrester is intended to absorb any undue surges on the system.

More costly forms of arresters are available such as the Peterson earth coil, which again is connected between neutral and earth or comb, electrolytic or oxide film arresters installed between each line and earth. The first consists of a coil of conductive material which, under normal conditions, provides a low reactance path to earth but, under the influence of a high frequency discharge, is highly inductive. The comb, electrolytic and oxide film arresters are similar to the arcing horns in that they operate on a recovery basis once the high frequency surge is discharged.

Unless an installation includes high voltage equipment it is not economically viable or practical to attempt to provide protection against such high level surges but, as previously mentioned, considerable increases above a low voltage system voltage can occur due, for example, to the switching operations on cage motors and circuit-breakers. However, provided that cables are adequately sized, this is more a problem for the switchgear designer than for the installer.

As the opening of a circuit-breaker does not always coincide with the normal current-zero, this leads to the phenomenon referred to as current-chopping, the effect of which, particularly in the case of light inductive currents, is to create over-voltages in both the circuit breaker and associated cables. This same phenomenon is used to practical effect in a number of electronic circuits to create high frequency supplies from the normal supply frequency more economically than by the use of rotary frequency-convertors. The technique, referred to as switched-mode power supply, utilizes the mains input current for a small period of time, measured in milliseconds, at the peak of the mains voltage waveform, resulting in a much larger effective (RMS) current than a similar sinusoidal one.

The increased use of switched-mode power supplies throughout commercial and industrial installations on equipment such as computers, data processors, starters for fluorescent luminaires and speed controllers for motors places an increased responsibility on the designer/installer to ensure that cable sizes are compatible with the requirements rather than the anticipated input power. This is of particular importance in installations consisting of three-phase and neutral distribution with numerous single-phase final circuits, as it may eliminate the possibility of using a reduced neutral (as permitted by the IEE Wiring Regulations where it is considered that a complete system may be reasonably balanced). It has, in fact, been suggested by computer experts that a neutral may, on large installations with numerous small electronic devices, carry far higher currents than the phase conductors (Fig. 17.9).

A building may have a lightning protection system installed, but this does not mean that everything within the building is protected. The lightning protection designed to BS 6651 is only protecting the fabric of the building and not the secondary effects of a lightning strike. In any event the building does not have to be struck by lightning for electronic equipment such as, telecommunication equipment, instrumentation systems or computers to be affected.

The electromagnetic field created by a lightning strike can induce a voltage into data transmission lines passing through the electromagnetic field. Also lightning strikes in proximity to the building will cause a rise in its electrical potential and hence a rise in potential with respect to true earth on the equipment inside the building. Modern technology uses data transmission lines to either transfer information or to control remote items of equipment such as, the control of a remote pump station in the

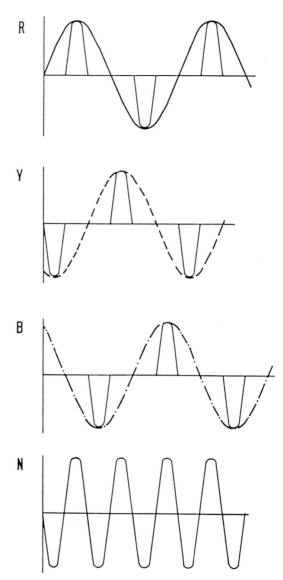

Fig. 17.9 Representation of resultant neutral current created on the mains by switched-mode power supply equipment. Balanced three-phase load.

water industry. Such transmission lines will therefore be running to a building having a much lower earth potential which will give rise to a transient overvoltage flowing to reduce the potential difference. This transient overvoltage will appear across components in equipment at either end of the data transmission line, which can damage electronic components severely.

Surge protectors should therefore be installed at the mains intake position to the building, and locally at important pieces of equipment. It may be worth considering protecting each item of equipment since plug-in surge protectors are now available at a very reasonable cost, which could well be less than the cost of time lost with equipment being out of action or the cost of the repair to the equipment, call-out charges could cost more than the surge protector.

The revised edition of BS 6651: 1992 when published will contain an appendix giving guidance on the protection of electronic systems, together with a risk assessment for electronic systems, methods of protection and how to select appropriate protectors.

Chapter 18
Fire Precautions and Protection

The advances in electrical technology, the discovery of better materials and a greater awareness of the possible dangers have all contributed to the greater safety of electrical installations and equipment. Despite this, the number, intensity and cost of both direct and indirect damage increases yearly, the actual losses recorded by the Association of British Insurers for 1991 was 1081.1 million pounds, this being the latest figure available. The figure quoted does not include any losses incurred by the government or by local authorities, so the actual losses will be well in excess of a billion pounds. Consequently, all concerned with the design and construction of new buildings must consider every aspect of fire prevention. Unfortunately too frequently, the blame for fires is placed on the associated electrics which, inevitably, suffer most damage at the heart of an incident and become an easy target for criticism. It has to be accepted that faulty installations can, and sometimes do, start a fire and, therefore, the installation engineer must ensure that the designs and equipment utilized are to the highest standards, and, particularly, that the electrics associated with fire prevention systems operate when required.

Fire precautions

All electrical equipment is a fire hazard as it generates heat intentionally or as a by-product of its function, and the degree of danger varies. Totally enclosed low-current devices are not normally a danger but, obviously, equipment such as heaters and semi-enclosed rotary equipment provide an increased level of hazard as there is the possibility that combustible materials in close proximity may be affected. Modern dry-type switchgear, although a potential hazard, is now manufactured to such high standards and so well enclosed that the danger of creating a fire is minimal, but the types containing mineral oil, and similar transformers, present greater problems.

General plant protection

Power-operated equipment, whether a simple luminaire, a solenoid or a large motor or transformer, provided that it is correctly rated for the purpose, is properly installed, maintained and used, provides no possibility of a fire being caused. In many cases, however, the installation engineer is not responsible for supplying the equipment, e.g. package boilers, air conditioning plant, refrigerators, etc., but the requirements for supply must be correctly ascertained to ensure that the circuitry — cables, protective devices and means of isolation — are adequate for the duty. When the equipment is included in the installation there is the greater responsibility and, as emphasized in the IEE Wiring Regulations, the highest standards are only achieved by the use of materials and equipment which conform to accepted standards.

Although insulation levels on electrical equipment are primarily related to the voltage and maximum temperatures at which it is to operate, the built-in safety factors provide some protection against abnormal, externally created rises in temperature. However, this is unlikely to give complete protection in the event of a fire; consequently, it cannot be regarded as any more than a first defence, and further precautions are required. In many cases such precautions are a statutory requirement or essential to comply with local by-laws; the Fire Precautions Act 1971, for example, covers buildings such as hotels and boarding houses, while the Factories Act at present covers factories, this may change in the future if the Fire Precautions (Places of Work) Regulations become law.

As far as houses are concerned the Smoke Detectors Act 1991 specifies that smoke detectors must be installed in all new dwellings. The smoke detector should conform to BS 5446 Part 1 and should be mains operated, but can also have a supplementary power supply to enable it to keep working for a period of time if the main power supply is lost.

Where the dwelling has more than one storey a self-contained smoke detector must be installed on each storey and the smoke detectors should be interconnected so that the activation of any one unit operates the alarm signal in all of them.

Electrical equipment for flameproof areas provides rather more protection than standard equipment, as it is specifically designed to prevent the transmission of sparks or flames from inside the FLP enclosures and to withstand explosion; obviously, the added protection works both ways.

In many cases it is possible, of course, to reduce the capability of electrical equipment to initiate a fire by, for example, building thermistors into rotating equipment which disconnect the units upon excessive rise of temperature. These features, again, should only be considered as first-line defence rather than as protection.

Liquid-filled equipment

Mineral-oil-filled transformers and switchgear are ostensibly the major sources of fire hazard in all installations, although experience has proved that, provided good maintenance procedures are applied, they are rarely the cause of a fire. However, it is common practice on large transformers to fit equipment such as Buchholz relays between the tank and conservator, which respond to surges in oil volume or the accumulation of gas, and temperature gauges or detectors, which give warning of unusual rises in temperature or trip the associated switchgear. The smaller oil-filled switchgear is usually provided with gas vents or the slightly more expensive bursting discs which are intended for the relief of internal pressure rather than to prevent fire.

Every installation of oil-filled equipment should be provided with means for safely retaining any oil spillage; generally sand or gravel-filled trenches adjacent to or surrounding the site are adequate and, if within buildings, there must be a substantial enclosure. An external substation should be remote from other buildings and securely fenced.

It is also advisable to provide suitable hand-operated fire fighting extinguishers for all substations and switchrooms and preferably, on indoor installations, to install automatic sytems as described in the following sections. Although other types of liquid-filled or air-insulated equipment are not considered to be as liable to fire as oil-filled, explosions can occur and, therefore, similar precautions should be taken.

Cables

Materials used for insulation and sheathing on the older types of power cables were more highly flammable than many of the current types but, as they were usually buried except at the switchgear, terminations, etc., they were unlikely to contribute to or maintain a fire. However, PVC cables are often installed above ground in high-rise buildings and factories, and this material is flammable and also produces toxic fumes. To reduce the possible danger, such cables should be installed wherever possible in fire retardant ducts or passageways or, otherwise, well clear of flammable materials. Where large numbers of cables are installed in enclosed walkways, consideration should be given to the use of automatic fire-fighting systems and vertical ducts should be provided with fire barriers at frequent intervals.

Developments in insulating and sheathing materials allow the production of cables which are capable of operating in high temperatures, are less flammable than PVC and reduce the creation of toxic fumes. As these cables are comparable in price to PVC and contribute far more to safety, they should be seriously considered for all indoor situations and, particularly, in areas which are known to be hazardous. Within this range of cables are the BICC Flambic types, which meet the flame propagation requirements of BS 6883: 1991, for power, lighting, control, instrumentation and communication purposes which are capable of operating at 90°C, as against the 70°C of PVC. The power cables, being subject to the 1000°C and the instrument cables to the 750°C three-hour test called for by the IEC 331 Standard. Additionally, four classes of sheathing material are available for armoured cables, thus enabling cables to be selected which have a neglible halogen acid gas emission.

Between PVC and Flambic cables is the cross-linked polyethylene (XLPE) insulated range which is also rated for 90°C but which, for general use, is usually sheathed with PVC; however, these are also available with alternative types of compound sheath which are more suitable to combat fire hazards.

Protection

Although, as mentioned, manufacturers include many features in their products to combat fire hazards, these do not in themselves provide fire-fighting capabilities, and therefore, the following methods should also be adopted.

Alarms

For relatively small installations where responsible personnel are available at all material times a simple alarm system may be sufficient, consisting only of audible or visual alarms throughout a building and operated from a single manned station or from manual remote break-glass units. With such a system a responsible person must always be in attendance, as notification of a fire can only be made by some means of communication by other personnel. If there are periods during which a building is completely unoccupied it is essential that an audible alarm should be of sufficient intensity to attract the attention of outsiders, and operation by some form of automatic detection system is required.

Various types of audible alarm are available – bells, klaxons, sirens, etc. – and any authorities involved are usually amenable to all of these, subject to audibility and the requirement that all sounders on a system shall be of the same type.

Except on the smallest installations, an alarm system only is not rec-ommended as it is completely dependent on the human factor; when used in conjunction with a detection system, it should be operated automatically from this.

Detection

Every precaution should be taken to detect a fire at the earliest possible moment, and it is unfortunate that few insurance companies encourage the installation of effective systems by offering realistic discounts on premiums such as those applied to sprinkler systems. However, it is obvious that early detection is beneficial to all concerned, and many types of system are easily available at considerably less than the direct and consequential cost of a fire.

The majority of systems installed are comprised of suitably located types of sensor, such as heat, rate-of-rise, optical, flame or ionization (smoke) detectors, wired back to an annunciator panel. This type of system is referred to as the 'point' system while the alternative, the 'line' type, consists of a pair of conductors insulated with heat sensitive material which, when affected by a rise in temperature, allows contact between the two wires and completes the alarm circuit. Upon reduction of temperature the insulation level must be restored and the circuit system re-activated.

To comply with good engineering practice and all applicable regulations, it is essential that the interconnecting cables are mechanically sound, protected against extraneous damage and completely segregated from all other systems. Although MICC cable meets all the requirements, other systems of wiring are not excluded by regulations and, therefore, subject to certain precautions, detection systems may be wired in PVC in steel or plastics conduit or in existing trunking.

Annunciator panels, except on small installations, should indicate the zones into which a building has been divided, give audible indication of an occurrence, confirm the availability of the power supply to the system, provide test facilities and, preferably, initiate alarms. A refinement is the provision of a two-stage alarm, which signals only in the immediate area affected by a fire in the first instance but provides a second tone throughout the building if the danger spreads further (Fig. 18.1).

Power supplies for these systems may be direct from the incoming supply, in which case they must be taken from as near as possible to the incoming cable, or through a battery/charger system. A direct supply is not recommended unless a reliable standby supply is also available. It is essential in both cases that the supply device for the detection system is

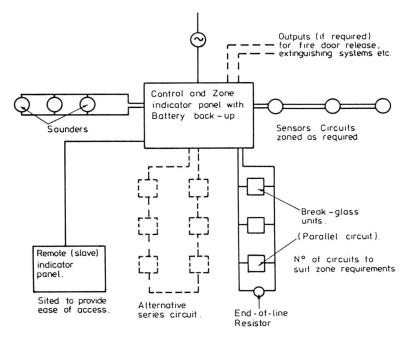

Fig. 18.1 Typical circuit layout for system comprising manual and automatic
   signalling equipment with central control and zone indicator panel.

clearly indicated. For detailed information, reference should be made to
the Code of Practice BS 5839: Part 1.

   Facilities are available, through the public telephone network and by
agreement with the fire service, to connect alarms and detection systems
directly to the fire brigade control centre so that, in the event of a system
operating, immediate call-out of the brigade is effected. In addition, some
of the larger manufacturers of alarm and detection equipment operate
their own control centres, to which signals may be directed and then
routed to the appropriate authority. These additional facilities entail
increased costs for the rental of dedicated telephone lines and the use of
the private control centre.

Protective systems

There are many different types of protective system available, ranging
from the manually operated cylinder-type fire extinguishers and hosepipes
for small or relatively unimportant premises to complex alarm, detection,

remote signalling and automatically operated fire-fighting installations. The equipment obtainable from major manufacturers is often capable of being easily extended as the need arises, which allows the systems to be tailored to the immediate requirements or capital available. It is again emphasized, however, that the cost of the most expensive system will almost certainly be less than that of a fire and the consequential results.

Hand appliances

Unless highly volatile materials are present, a fire tackled in the early stages can be prevented from turning into a major disaster, and many types of portable pressurized extinguishers are available for this purpose. These are particularly useful in confined areas, such as substations and switchrooms, or areas not covered by an automatic system. Where there is the slightest possibility of electrical equipment being affected, however, only carbon dioxide (CO_2), and certain types of foam extinguisher should be used, i.e. not water-based contents. Such equipment should be clearly identifiable and located in easily accessible positions on escape routes. CO_2 is a health hazard, however, and must not be used in such a manner that personnel may be trapped away from the exits etc. Some installations currently use Halon, but manufacture of this product ceases at the end of 1993.

It is strongly advised that specified personnel should be authorized to use portable equipment and fully trained in the correct methods of use and application.

Sprinkler systems

There are two basic types of sprinkler system, generally referred to as 'wet' and 'dry'. In the wet system, all pipework is continuously charged with water under pressure and, therefore, rises and falls in pipework are, to some extent, immaterial. Obviously, the head on a riser must always be considerably less than the system pressure. The dry system, when not in operation, is pressurized with compressed air, and facilities are provided for draining down after use and, therefore, falls in the pipework must be avoided, even to the sprinkler heads. For this reason, the heads are for inverted operation mounted above the system pipework. Where low temperatures are present the dry system is to be preferred, as it eliminates the possibility of a freeze-up in the pipework.

A sprinkler system requires an immediate and adequate supply of water at a suitable pressure provided, preferably, from its own pumping station or from a storage tank to avoid nonavailability of the system through loss of a mains water supply. Additionally, with the dry system an automatically operated compressor is required to maintain the air pressure.

For the electrical designer/installer the disadvantages of a sprinkler system are that water extends any damage suffered by unenclosed electrical equipment from a fire and the pipework layout often prevents uniformity in the lighting installation. Wherever possible, therefore, luminaires should be mounted at a higher level than sprinkler heads, and all other equipment in a sprinkler zone must be suitably protected against the effects of water. It is also essential that electrically operated plant and control gear associated with a sprinkler system is supplied with power from the most secure source.

A simple alarm system must have some means of energizing the sounders, i.e. break-glass units, but it is also of advantage to include these in detection, sprinkler, and CO_2 installations. In the majority of installations it is, in fact, an essential requirement in order to comply with statutory or enforcement authority rules and regulations.

Reference should be made to BS 5839, BS 5445 and BS 5446, the last being specifically applicable to residential premises. Also, many countries have their own certifying bodies for fire detection and fighting equipment, such as the UK Loss Prevention Council, and major insurance companies insist upon the use of such approved equipment.

Similar sprinkler heads are used on both wet and dry systems, each one containing a sealing device, usually a glass bulb filled with a temperature-sensitive material. This disintegrates at a preset temperature, allowing the sealing plug to be ejected by the water (wet system) or air (dry system) pressure (Fig. 18.2). The resultant drop in pressure in the pipe network activates a pressure-sensitive valve at the pumphouse which, in turn, starts the system pumps. In the dry system the pump and compressor drives are suitably interlocked to prevent both operating simultaneously.

It will be clear that, in the above systems, only the heads affected by the heat of a fire will come into operation, and the discharge of water is, therefore localized. If a full drenching system is required, open heads are available but alternative means of detecting heat must be incorporated in the control system. It is possible to accomplish this by linking the control system to a fire alarm and/or detection installation.

In most cases a sprinkler system must conform to the rules, in the UK, of the Loss Prevention Council, i.e. LPC rules, and the major insurance companies allow generous discounts on premiums which, as mentioned in an earlier section, are not equalled by alarm and detection systems.

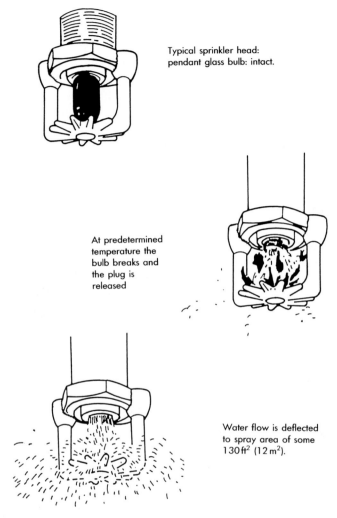

Typical sprinkler head:
pendant glass bulb: intact.

At predetermined
temperature the
bulb breaks and
the plug is
released

Water flow is deflected
to spray area of some
130 ft² (12 m²).

Fig. 18.2 Mode of operation of sprinkler head. (Courtesy of Fireproof
Engineering Ltd.)

Carbon dioxide systems

There is similarity between sprinkler systems and a carbon dioxide (CO_2)
installation in that both require pipework and distribution heads connected
to storage tanks containing the extinguishing media. The CO_2 storage,
however, requires far less space than pumps and water storage, as the
media is self-ejecting when control valves are opened. CO_2, which is a gas
at normal temperature and pressure, is supplied as a liquid under pressure in

easily handled vessels which are simply coupled into the system and, due to the relatively small size, more easily accommodated. Consequently, large buildings may be covered economically by the installation of a number of completely independent stations, each one sized according to the area to be protected.

When the storage cylinders have been coupled into the system their manual retaining valves are opened, leaving the pressure maintained by a master automatic valve. The most simple arrangement for opening the master valve consists of a weighted handle on the valve, held in the closed position by a strainwire (or network of wires) linked to a thermal coupling. An unusual rise in temperature melts the coupling, which releases the strainwire and allows the valve to open under the influence of the weight. Release mechanisms are also available which utilize solenoid-operated master valves, detector heads and electronic control systems which are more adaptable to modern requirements than the simple mechanical system.

A CO_2 system has the great advantage over a sprinkler system that there is no consequential damage although, as it is an asphyxiating gas, when installed in confined areas such as substations and switchrooms precautions must be taken to ensure that it cannot be triggered off inadvertently by the inclusion of a lock-off in the control system.

It must be emphasized that precautions must be taken with CO_2 since it is ,toxic but, again, it causes no consequential damage and can safely be used to protect computer and similar highly sensitive installations.

Types of detector

Heat	Responds only when a given preset temperature is attained from any cause.
Rate-of-rise	Responds to the speed of temperature increase which, normally, is more rapid in the case of fire than from natural causes.
Optical	Responds to the obscuring of a beam of light, e.g. emitter and photo-cell.
Flame	Responds to an increase of the illumination level, e.g. infrared detectors.
Ionization (smoke)	Responds to a change in electrical characteristics, i.e. capacitance in the detector caused by the entry of smoke or fumes.
Smoke (visible)	Responds to light-scatter caused by smoke particles, e.g. photo-sensitive detectors.

Chapter 19

Installation, Inspection and Testing

It is generally accepted that systems operating at voltages above 50 V a.c. and 120 V d.c. are potentially lethal and, while some countries such as the USA have attempted to reduce the possible danger by supplying the end-user at voltages as low as 110 V, in the UK the standard is 240 V single and 415 V three-phase. Should a consumer prefer to use a lower voltage for his installation he may do so, but the cost-involvement is his completely. It is emphasized, however, that in some instances, such as on construction sites, there is a strong recommendation by the UK Health and Safety Executive to provide lower voltages for construction purposes. In quarries lower voltages are mandatory for certain items such as portable equipment.

From the above it is obvious that the maximum degree of safety is only obtained by ensuring that all installation work is carried out to the highest standards of workmanship and materials, and that adequate procedures are followed for inspection, testing and maintenance.

Standard of installation

As emphasized throughout this book, the most commonly used guide to good installation practice, although it carries no statutory obligations, is the IEE Wiring Regulations for Electrical Installations. However, there are several Acts in existence which cover special industries (these are listed in Appendix 1) and a number of the larger UK organizations issue their own standards, generally orientated towards specific processes.

Probably the greatest difference between statutory documents and the IEE Wiring Regulations is that the former, generally, stipulate the final requirements while the Regulations give the methods to be adopted to achieve them, although they do not exclude alternative sound engineering practices. In Chapter 13 of the IEE Wiring Regulations, Regulations 130−01 to 130−10 are comparable to statutory regulations, the relationship with which, is explained in Regulation 120−02−01.

The IEE Wiring Regulations are only concerned with the fixed wiring of the installation, in this respect, the Health and Safety Executive, in their statement on page vii of the IEE Wiring Regulations, consider that compliance with the IEE Wiring Regulations is likely to comply with the relevant aspects of the Electricity at Work Regulations 1989. To the designer, however, Parts 3, 4, 5 and 6 together with the appendices are of the major importance and must be studied in detail.

Safety aspects

The main purpose of an electrical installation is to provide the user with his requirements while giving the maximum protection to people, animals (where necessary) and property, from shock, burns, injury from mechanical movement of electrically operated equipment and fire. In addition to the possible dangers directly associated with an installation, however, an adequate design should also take into consideration other safety require-ments, including adequate illumination under both normal and abnormal conditions (emergency lighting), detection and warning in the event of a fire (fire alarm and detection system), and protection against the effects of adverse weather conditions (lightning protection), while in some cases, particularly continuous process industries, the aspects of safety will also encompass standby electricity supplies.

Reference should also be made to Regulation 514−09 which emphasizes the requirement for comprehensive drawings and/or schedules incorpor-ating all the details of an installation including locations.

Testing to regulations

Before an installation is commissioned it is essential that a full range of tests is carried out as described in Part 7 of the IEE Wiring Regulations, but it should be noted that, before this is commenced, a thorough physical inspection should be effected. Previous editions of the Regulations included this requirement and to be effective, inspection should be an ongoing process from the commencement of a project, leaving only the obvious items such as the fitting of trunking and conduit-box lids, the presence of danger and warning notices, diagrams, instructions and similar information to be checked at the end. The applicable regulations are 711−01 and 712−01. Section 712 gives detailed lists of items requiring checking during the inspection.

The subject of testing is covered by Part 7 with the standard methods required to comply with the regulations detailed in Section 713.

An important point arising from the requirements for testing the conti-nuity of both live and protective conductors and the earth loop impadence

test is that measurements must now be much more precise and, consequently, the installer has less freedom than formerly with regard to the
relocation of cables and equipment if such relocation involves increases in
route lengths. Any proposed alterations from the original design must,
therefore, be given close consideration before being effected.

Prior to a supply being available it is necessary to obtain from the
regional electricity company the fault level and the external impedance
(Z_E) at the origin of the installation, as these factors govern the characteristics of the proposed protective devices and the design parameters of the
installation cabling. Unfortunately, the figures provided may not be
completely accurate and, therefore, it is strongly advised that as soon as
the service has been commissioned the accuracy be checked by means of
equipment such as the Merlin Electrical Distribution and Test's IMPSC
instrument (Fig. 19.1), which was designed specifically to meet the requirements of the 15th Edition by Dorman and Smith, and with a wide-scale
earth loop impedance tester.

If the test figures depart greatly from those provided, it may mean, at
the worst, that the whole installation design must be modified.

The three organizations associated with electrical installation work are:
the Electrical Contractors Association (ECA) for England and Wales, the
Electrical Contractors Association Scotland (ECAS) and the National
Inspection Council for Electrical Installation Contracting (NICEIC). The
ECA and the ECAS check the quality of work before allowing a company
to become a member of their organization. The NICEIC carries out an
initial inspection and test before allowing a company to become a member
and then undertakes a sample inspection and test of the work undertaken
by each of its members annually, to ensure that the correct standard of
workmanship is maintained. The responsibility of the above organizations
is limited since they can only exercise control over their own members.
Even this control is limited in some respects, in that failure of a company
to maintain the high standards set by these organizations only results in
the company's membership of the organization being terminated. This
can, of course, affect the company's trading since many organizations
insist that a contracting company is a member of one of the above
organizations. Membership is therefore, for the majority of members,
highly prized which keeps up the standard of workmanship.

The Health and Safety Executive has the right by law to enter any
premises where work is carried out and demand rectification of any
contravention of statutory regulations; this can take the form of an
improvement notice or in serious cases a prohibition notice, which can
immediately stop work continuing.

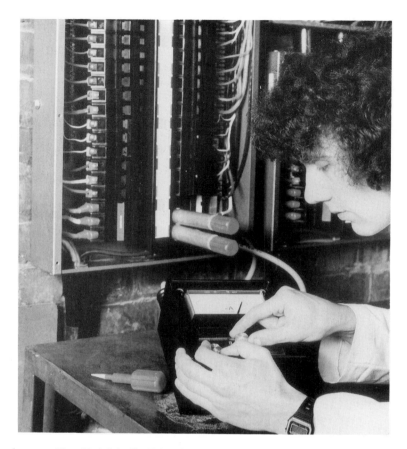

Fig. 19.1 Merlin IMPSC short-circuit test instrument.

In addition to the above organizations, many local authorities operate inspection departments, but these are usually directed at the effectiveness of systems rather than the technical aspects, e.g., emergency lighting and fire alarm systems.

Testing procedure and instruments

Continuity of protective conductors

The physical inspection should have verified that every bonding conductor's and protective conductor's connections are sound. They have now to be tested to verify they are electrically sound; a suitable instrument is shown in Fig. 19.2.

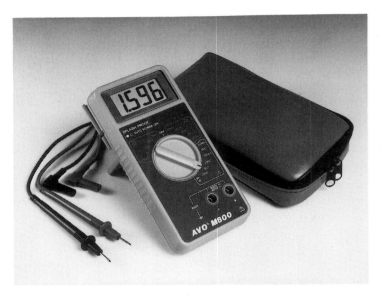

Fig. 19.2 Avo International M800 digital volt ohmmeter.

The test is made using a low resistance ohmmeter that has a minimum short-circuit test current of 20 mA with an open circuit test voltage between 3 V and 24 V. These values come from the IEE Guidance notes, but it is anticipated that in January 1998 these values will be changed to the CENELEC and International Wiring Regulations values given in HD 384 and IEC 364. At present HD 384 and IEC 364 specify a short-circuit test current of 200 mA and an open circuit test voltage between 4 and 24 V. Fortunately there are continuity test instruments on the market that do comply with HD 384.

For wiring such as, twin and CPC cable, the phase conductor can be connected to the earth bar in the distribution board and tests between phase and earth made at outlets in the circuit. A reading should be taken at the end of each circuit and recorded; this reading being added to the external earth loop impedance Z_E and then checked against the Regulations for compliance with protection against indirect contact. Do not forget, however, that the values of Z_S given in the regulations are at the design temperature.

An alternative method of checking the continuity of protective conductors is to connect one test lead to the main earthing terminal and the other test lead to protective conductors at various points in the circuit. In both types of test, if the resistance is to be used, the resistance of the test leads must be deducted from the readings obtained.

Where bonding conductors are installed to comply with Regulation

413−02−15 or Part 6 of the regulations, it shall be verified that the bonding conductor's resistance complies with the appropriate formula given in Regulation 413−02−16 or in Part 6.

Where the protective conductors are made from ferrous materials such as, steel conduit, trunking etc., the integrity of the conductor should have been verified during the physical inspection. It is more appropriate that such items are inspected as the installation proceeds.

The test is made with a low resistance ohmmeter to verify the integrity of the ferrous protective conductor. If there is some doubt about the integrity, the tests are repeated using an earth loop impedance tester. If, after this test, some doubt is felt about the integrity of the protective conductor, the test can be repeated using an a.c. high current low-impedance ohmmeter. The ohmmeter should have a maximum output voltage of 50 V. The test is made by passing a current, subject to a maximum of 25 A, of 1.5 times the design current of the circuit.

Continuity of ring final circuits

Using a low resistance ohmmeter the continuity of each conductor in the ring circuit is made to ensure that no interconnections have been made.

There are several ways of testing the continuity of a ring circuit, but the method suggested in the Guide to the IEE Wiring Regulations is as follows. At the origin of the ring circuit the phase conductor of one leg is connected to the neutral conductor of the other leg and a reading obtained between the remaining phase and neutral conductor. This reading proves that the ring is continuous (see Fig. 19.3).

Next the phase and neutral conductor are disconnected from the test instrument and connected together. A test is then made at each socket-outlet on the ring. If there are no interconnections the reading obtained at each socket-outlet should be substantially the same. If the conductors are wrongly connected together, i.e., they are from the same leg of the ring, the reading at each socket-outlet will be different; the readings will increase up to a maximum at the midpoint of the ring.

Where twin and CPC cable has been used the test is repeated using the phase conductor and the CPC. The reading obtained at the midpoint of the ring will be the phase earth loop impedance, Z_{inst} i.e., $R_1 + R_2$, for the ring circuit. Adding this value to Z_E for the ring circuit will give Z_S, which can be checked for compliance with the regulations, after adjustment for temperature.

Insulation resistance

The instruments used for insulation testing must now be capable of supplying the test voltage with a 1 mA load. The test voltage applied

Continuity test on final ring main circuit

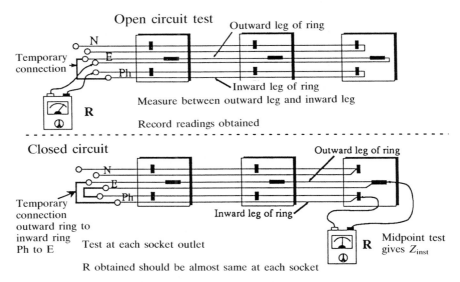

Fig. 19.3 Testing ring circuit at each socket outlet.
(Taken from *Handbook on the IEE Wiring Regulations* by T.E. Marks.)

being dependent upon the type of circuit being tested, a suitable instrument is illustrated in Fig. 19.4.

Where the circuit is an extra-low voltage circuit (0 to 50 V) supplied from a safety isolating transformer complying with BS 3535, and has been installed in accordance with the regulations for class II equipment, the test voltage shall be 250 V d.c. The minimum insulation resistance allowed being $0.25\,M\Omega$.

For low voltage circuits operating at a voltage between 50 V and 500 V the test voltage is 500 V d.c., and the minimum insulation resistance allowed is $0.5\,M\Omega$. For circuits between 500 V and 1000 V the test voltage is 1000 V d.c., the minimum insulation resistance allowed being $1.0\,M\Omega$. The test between SELV circuits and associated LV circuits is 500 V d.c., with the minimum insulation resistance allowed being $5\,M\Omega$. An additional withstand test of 3750 V RMS a.c. for 1 min has to be made where called for in British Standards.

The object of the above tests is to verify that the insulations of live conductors, accessories and protective conductors are satisfactory and comply with the IEE Wiring Regulations. Before any tests are made any items that may be damaged by the tests, such as electronic equipment, should be disconnected. Similarly, any equipment such as capacitors that

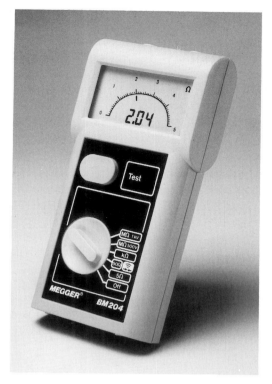

Fig. 19.4 Avo International BM204 insulation/continuity tester.

are permanently connected between phase and earth should be disconnected.

The insulation resistance has to be measured between each live conductor and earth and between live conductors. For this latter test it will be necessary to disconnect equipment that has a permanent connection between live conductors for example, immersion heaters.

Where equipment that has exposed conductive parts has been disconnected, an individual test has to be made between all live parts and exposed conductive parts, to verify the insulation resistance complies with the requirements of the appropriate British Standard. Where no such standard exists the minimum insulation value allowed is $0.5\,M\Omega$.

Site applied insulation

Where insulation is applied on site to protect against direct contact, it has to be subjected to a test voltage equivalent to that specified in the British

Standard for similar type tested equipment, to verify the insulation will not break down and that a flashover will not occur.

Where insulation is applied on site to protect against indirect contact, it shall be subjected to similar tests to those given above, to verify the enclosure is capable of withstanding the test without breakdown or flashover. Additionally, it shall be tested to ensure that the protection afforded is not less than IP2X.

Protection by separation of circuits

Where protection against electric shock is provided, either by SELV or by electrical separation, tests are made to ensure the installation complies with the appropriate regulations given in the 16th Edition of the IEE Wiring Regulations.

Protection by barrier or enclosure

Where a barrier or enclosure is provided during erection to protect against direct contact, in accordance with Regulation 412–03, a test has to be made to ensure that the degree of protection is not less than IP4X for readily accessible horizontal top surfaces and IP2X for all other surfaces.

Insulation of nonconducting floors and walls

These tests should be carried out by a competent person trained in the use of high-voltage test equipment. A detailed procedure is available from the UK health authority which is applied to such areas as hospital theatres.

The inspection and test involves verifying: that a person cannot come into contact simultaneously with two exposed conductive parts or with an exposed conductive part and an extraneous conductive part, that there are no protective conductors in the location and that socket-outlets do not contain an earthing contact.

A test is made between three widely separated points on each surface, one point of which is not less than 1 m and not more than 1.2 m from an extraneous conductive part, and the main protective conductor for the installation. The resistance should not be less than $50\,k\Omega$ where U_0 does not exceed 500 V and $100\,k\Omega$ where U_0 is between 500 V and 1000 V. If these values cannot be achieved at any point the floors and walls are deemed to be extraneous conductive parts.

Where extraneous conductive parts are insulated so as to avoid simultaneous contact with exposed conductive parts they should have an insulation resistance of $0.5\,\text{M}\Omega$ when tested at $500\,\text{V}$ d.c., and should be able to withstand a test voltage of at least $2\,\text{kV}$ RMS, a.c. and in normal use not pass a leakage current in excess of $1\,\text{mA}$.

Polarity

In the majority of cases polarity cannot be determined by visual inspection since conductor colours can be changed at joints in the system. A test is therefore essential to ensure that every fuse, circuit breaker and control switch, as well as the centre-contact of bayonet or Edison screw lamp holders, are connected to the phase conductor and that socket-outlets have been correctly connected.

Earth fault loop impedance

The test equipment works by connecting a resistance of about $10\,\Omega$ between the phase and earth conductor, so that a current between $20\,\text{A}$ and $25\,\text{A}$ flows through the circuit. The instrument's meter is connected across the resistor; the meter being calibrated to take into account the voltage drop due to the resistor when the fault current flows.

A dangerous occurrence can occur whilst testing if the earth loop impedance is high and the time the current is flowing is not limited, since the protective conductors and hence the exposed conductive parts of the installation can rise to mains potential. It is for this reason that modern earth loop impedance testers incorporate a timer disconnecting the circuit within $40\,\text{ms}$.

Two tests are required, the first test being at the origin of the installation to determine the value of Z_E (see Fig. 19.5). For this test the equipotential bonding conductors are disconnected from the main earth terminal and the test is made between the incoming phase conductor and the main earth terminal. For safety reasons the main CPC going from the main earth terminal to the switch gear, or the protective conductors connected to the main earth terminal, should be disconnected, otherwise there is the danger of raising the installation's exposed conductive parts up to mains potential. Additionally, the phase conductor supply to the instrument should be through a cartridge fuse which will disconnect the circuit quickly in the event of a short-circuit. If the test is made close to the mains intake the resistance of the tails to the switch gear should not have much influence upon the readings obtained.

The second test should be made at the end of each circuit, but tests at intermediate points can be made if desired. For this test the equipotential

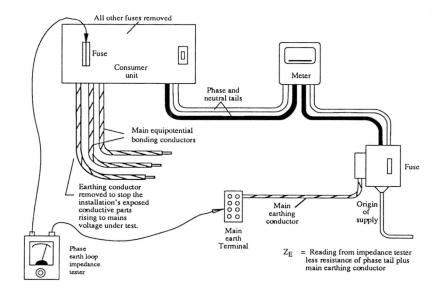

Fig. 19.5 Determination of Z_E at the origin of the installation.
(Taken from *Handbook on the IEE Wiring Regulations* by T.E. Marks)

bonding conductors are reconnected (together with the main CPC and any other CPCs that were disconnected) since they are involved in the phase earth loop. A typical instrument for these tests is illustrated in Fig. 19.6.

All test results should be recorded but the results of the second test should have the value of the first test, or the declared Z_E provided by the electricity supplier subtracted from their value. The resultant earth loop impedance for each circuit is then adjusted to convert the resistance from the temperature at the time of the tests up to the design temperature for the circuit. This final figure is then compared with the maximum Z_S allowed for the circuit to verify that protection against indirect contact has been provided.

Where Table 41C has been used as the method of protection against indirect contact, for a circuit or for a distribution board having mixed circuits, the impedance of the protective conductor has to be measured to verify compliance with Table 41C.

Earth electrode resistance

Earth electrodes have to be installed for TT and IT systems and on large installations for earthing the star point of the associated transformer.

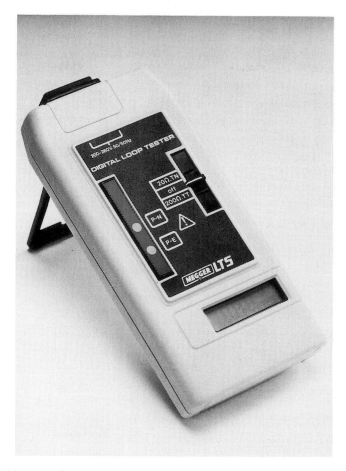

Fig. 19.6 Avo International digital phase earth loop impedance tester.

Earth electrode tests are usually carried out using a null balance earth tester, which passes a current down the earth rod through the ground and back to the instrument through a test probe inserted into the ground. At the midpoint between the earth rod and the test probe a second test probe is inserted into the ground which measures the potential between the earth rod and the midpoint test probe. This potential is applied to a Wheatstone bridge, which contains adjustable resistors, in the instrument. The resistors are adjusted until no voltage appears across the meter connected to the bridge, i.e., there is a null balance. The value indicated by the resistors is taken to be the earth electrode resistance. The midpoint probe is moved some 3 m nearer the earth electrode and then 3 m further away and the test repeated at each position. Providing there is an insignificant difference between the three readings the average of the three readings is taken as the resistance of the earth electrode.

If the needle of the meter in the instrument fluctuates it is probably due to stray currents in the ground; this problem can be overcome by varying the speed by which the generator handle on the instrument is rotated.

Only a competent person should be allowed to carry out tests, especially for existing installations, since care must be taken due to possible fault currents causing voltage gradients round the earth electrode.

One method of testing an earth electrode, given in the IEE guide to inspection and testing for TT installations, is to use a phase earth loop impedance tester. This will give a result which is probably suitable for such installations, but testing at mains voltage must be a possible source of danger to the person carrying out the tests and other personnel on the site. If this method is to be adopted it should be accompanied by a work schedule and in some cases a permit-to-work system.

Residual current devices

Before carrying out any tests the trip mechanism on the RCD should be operated to ensure that the RCD is not faulty and any load connected to the RCD should be switched off.

The RCD tester is usually equipped with a switch which sets the short-circuit current at 50% and 100% of the RCD rating. Additionally, the fault current can be switched to 150 mA (see Fig. 19.7).

The first test is made between the phase and protective conductor, preferably at the end of the circuit so that the maximum resistance is in circuit with the RCD tester. The first test is made with the instrument set at 50% of the rated tripping current; the RCD should not trip with the test applied for 2 s. The second test is made with the instrument set at 100% of the rated tripping current; this time the RCD should trip in less than 200 ms.

Where RCDs incorporate a time delay the time of the above tests is increased by the time delay of the RCD. With such RCDs it is important that the CPC does not rise more than 50 V above the earth potential ($I_{\Delta n} \times R_2$). It is therefore recommended that the test does not exceed 2 s.

Where an RCD is used for supplementary protection against direct contact the test is made at 150 mA; in this case the RCD should trip within 40 ms and the test must not be applied for more than 50 ms.

On completion of the tests the RCD should again be checked by using the test button on the RCD. Some instruments are provided with a negative half cycle switch so that the test can be made in each half cycle.

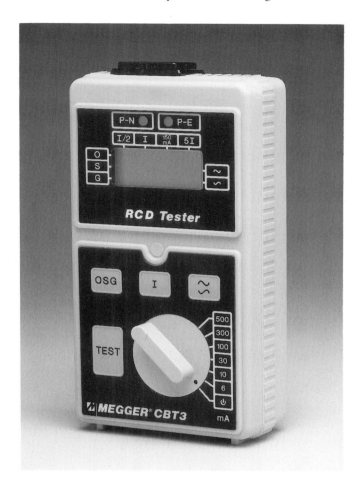

Fig. 19.7 Avo International RCD tester.

Testing to Standards and Codes of Practice

In the UK the most-used documents for testing procedures are those issued by the British Standards Institution, which is an independent body funded by the Government and industry, and although, as with the IEE Wiring Regulations, the publications are not statutory, they are accepted by the appropriate testing authorities as official guidelines. This is not unrealistic, as all British Standards are produced by committees representing applicable government departments (such as the Health and Safety Executive), enforcing authorities, professional institutions and industry (the latter being drawn only from trade associations and similar to avoid manufacturing bias). A great deal of effort is in progress towards the harmonizing of British, international and European Standards, which reduces many of the problems that existed between home-produced and imported equipment.

As it is the responsibility of the manufacturer to ensure that his products comply with the appropriate standards, the specifier has no control over this aspect except with regard to selection. However, Section 511 of the IEE Wiring Regulations does place the onus on the user to ensure that, wherever possible, equipment complying with a British Standard or Harmonized European Standard is installed.

Where equipment complying with a foreign standard based on an IEC Standard is used, the designer or specifier has to verify that any variance between that standard and the British or Harmonized European Standard, will be just as safe as if the equipment was manufactured to a British Standard.

Where equipment is used that is not covered by either a British or Harmonized European Standard the designer or specifier has to verify that the equipment is no less safe than that required by the IEE Wiring Regulations.

One area of testing that may present difficulties for the installer is with regard to the construction of switchboards, in that the IEE Wiring Regulations permit site-assembled units to be installed but demand that such assemblies be subjected to the same rigorous tests as factory-assembled switchgear, a requirement that almost certainly is beyond the capability of the normal installer as it entails the use of specialized and expensive equipment. However, the range of switchgear available and the relatively high cost of labour for site-assembly work virtually excludes this operation. In those instances where small, simple switchgear assemblies are required, they are normally constructed from individual items such as isolators, switch-fuses, MCBS, MCCBS, distribution boards, etc., all of which have been subjected separately to the necessary tests during manufacture and, therefore, the problem is eased.

In the UK the guarantor of quality for switchgear is the Association of Short-Circuit Testing Authorities which, subject to equipment proving satisfactory over an extensive range of tests, issues an ASTA certificate to the manufacturer. The required procedure for certification is that the prototype along with the manufacturing drawings are submitted to the Association. The Association then carries out the tests and issues approval subject to all other items being manufactured strictly in accordance with the submitted drawings and prototype. This means that other items manufactured are subjected to a less onerous routine testing. The applicable tests are fully detailed in appropriate British Standards which are generally aligned with international standards.

Where a company manufactures a product they are at liberty to put a label on it or state on the product that it has been manufactured to, or complies with, a British Standard number. This is only an opinion expressed by the manufacturer, who if he is wrong, and the product does not conform to the quoted British Standard, is liable to prosecution by the Trading Standards Office.

Where a company wishes to display the British Standard Kitemark which independently confirms the product is made to the appropriate British Standard, the company is required to make an application to the British Standard Institution, Quality Assurance. BSI Quality Assurance will then visit the factory to see the product being manufactured; will make a BS 5750 assessment for that product area to ensure that the standard will be maintained and will then take products away with them and subject them to tests at the BSI Test House at Hemel Hempstead. If all tests and the assessment are satisfactory the manufacturer will be given a licence to use the Kitemark. The licence is renewable annually providing BSI's continuing assessments of the quality system and the product continue to meet the standards.

A different level of quality assurance is provided by BS 5750 Quality System Assessment and Registration. BS 5750 defines the requirements for procedures to ensure specifications for products or services can be complied with. BSI Registered Firms are subject to continuing assessments by BSI to ensure that their quality systems remain effective. Products are not independently sampled and tested as they are with the Kitemark.

Similar organizations to ASTA exist in other countries, such as KEMA in Holland, which are authorized to test to both their own national and British Standards, and to issue certificates which are equally acceptable in the UK.

The IEE Wiring Regulations only specify that materials used should be manufactured to British or Harmonized European Standards and not that approved materials should be used.

Records and routine testing

Reference was previously made to Regulation 514–09 which requires the provision of diagrams, charts, tables or equivalent forms of information. These documents should include information on the type and composition of circuits, means of identification of protective, isolating and switching devices, and the methods used for designing an installation to comply with Regulation 413−02−04, i.e. the characteristics of protective devices, earthing arrangements, etc. It would appear from this that, if a design has been correctly accomplished, all of the major information will be available in the design drawings and/or schedules, leaving only such items as changes of requirements during construction to be recorded together with the provision of the appropriate Completion and Inspection Certificates as indicated in Appendix 6. Recommended symbols are given in BS 3939.

Of further importance to the life and continued effectiveness of an installation is the application of a routine testing procedure, and reference is made to this in Chapter 34 of the Regulations. Although this is obviously the responsibility of the user, there is an onus placed upon the designer/ installer for an assessment to be made of the frequency and quality of maintenance and, from this, the applicable inspection and testing periodicity. A recommended form of notice is given in Regulation 514−12−01. The recommended periods for periodic inspection and testing are given in the IEE Guide to Inspection and Testing. In many cases it will be found that the BSI Codes of Practice provide rather more detailed information on inspection periods applicable to the subject of the codes. Since the advent of the Electricity at Work Regulations the Health and Safety Executive is now beginning to specify periods of inspection and testing for various industries.

It is unfortunate that, frequently, due to the type of operation being carried out in a building, it is difficult or even impossible to adhere strictly to recommended periods or to carry out all the necessary operations. It is stressed, therefore, that the quoted periods are maximum ones and, for the benefit of the user, routine testing must be as comprehensive as possible, even to the extent of shutting down processes or operations completely on a planned basis rather than suffering the disruption of an unanticipated breakdown. There is a legal obligation to carry out periodic inspection and testing by Regulation 4(2) of the Electricity at Work Regulations 1989.

Inspection

The need for inspection procedures has already been mentioned in this chapter, but it is necessary to discuss the subject more fully as adequate inspection entails a complete awareness of the requirements of the preceding subjects: standards of installation, safety aspects and testing to both regulations and applicable standards.

The IEE Guide to Inspection and Testing provides a check list of items required to comply with Section 712 where these are relevant to an installation, but these should be taken as the minimum requirements as, obviously, all installations differ and, consequently, methods of installation or equipment may be introduced by the designer/installer that were not foreseen when the 16th Edition was issued. Additionally, other methods of complying with the regulations are not precluded, provided that they produce the same effects as those intended.

Regulation 130−01−01 refers to good workmanship and proper materials and, although the first may be obtained by training and experience, errors may arise which, without adequate inspection, could give rise to problems or serious faults at a later stage. For example, omission of the removal of burrs from the cut ends of steel conduits could lead to insulation being stripped from cables as they are being drawn in, leading to earth faults or short-circuits. Whilst for small companies with few contracts it would be unrealistic to employ personnel purely for inspection purposes, larger companies should have independent inspection departments responsible solely to the directors of the company.

Such departments should be responsible for routine inspections for quality during the progress of the contract and for the final inspection and testing when the contract is complete. One further duty that could be given to the inspection department would be that of ensuring the as-installed drawings and maintenance manuals were correctly correlated and handed over to the user of the installation.

In addition to site inspections, there is the further requirement for the use of proper materials referred to which, to comply with Regulation 712−01−02, should, when possible, be in accordance with British or equivalent standards. Initially, this responsibility is that of the installation designer who, from his knowledge of the requirements of the end-user and of the purpose and conditions under which equipment and materials are to be installed, prepares the specification for a project. There are instances, of course, when a suitable standard does not exist, as in the case of purpose-made luminaires. However, it will be seen from Appendix 1 of the IEE Wiring Regulations that, although this is not a complete list of all British Standards applicable to electrical equipment, it covers practically all of the major items; many of the smaller components, such as relays, contactors, ballasts, capacitors, etc. are covered by British Standards. Consequently, where a British Standard does not exist for a complete item of equipment, it will often be the case that components are covered, and inspection should ensure that only such approved items are incorporated in specialized equipment.

Safety aspects dictate that an installation shall provide the user with his requirements without danger and, also, that the equipment etc., shall be properly protected against external influences and, although chapter 32 of the 16th Edition is, at present, only reserved for this subject, reference should be made to Appendix 5, an alphanumeric schedule based on IEC Publication 364 which classifies and codes many of the anticipated influences. Additionally, BS 5490 Specification for degrees of protection by enclosures, includes an index of protection (IP) code which, at present, covers the ingress of solid bodies and liquids; it is anticipated that the IP will be extended to include protection against mechanical damage by the addition of a third numeral (Table 19.1).

It would appear, therefore, that the intent of the IEE Wiring Regulations is, in addition to site inspections, to ensure that an adequate stock control procedure is effected which will satisfy the installer not only that materials and equipment are available as and when required but that they also comply with both technical and physical requirements.

The requirements for site inspection, as referred to previously, differ for numerous reasons, but there are many aspects of installations which are common to all. It is strongly recommended, therefore, that every organization engaged in installation work should prepare a standard schedule such as Table 19.2 for inspection staff, with provision for the inclusion of any additional items necessitated by a particular project.

Table 19.1 Index of protection for enclosures.*

First numeral		Second numeral	
(a) Protection of persons against contact with live or moving parts inside enclosure		Protection of equipment against ingress of liquid	
(b) Protection of equipment against ingress of solid bodies			
No./symbol	Degree of protection	No./symbol	Degree of protection
0	(a) No protection.	0	No protection.
	(b) No protection.		
1	(a) Protection against accidental or inadvertent contact by a large surface of the body, e.g. hand, but not against deliberate access.	1	Protection against drops of water. Drops of water falling on enclosure shall have no harmful effect.
	(b) Protection against ingress of large solid objects <50 mm diameter.		
2	(a) Protection against contact by standard finger.	2	Drip proof: Protection against drops of liquid. Drops of falling liquid shall have no harmful effect when the enclosure is tilted at any angle up to 15° from the vertical.
	(b) Protection against ingress of medium size bodies >12.5 mm diameter ≤80 mm length.		
3	(a) Protection against contact by tools, wires or suchlike more than 2.5 mm thick.	3	Rain proof: Water falling as rain at any angle up to 60° from vertical shall have no harmful effect.
	(b) Protection against ingress of small solid bodies.		
4	(a) As 3 above but against contact by tools, wires or the like, more than 1.0 mm thick.	4	Splash proof: Liquid splashed from any direction shall have no harmful effect.
	(b) Protection against ingress of small foreign bodies.		
5	(a) Complete protection against contact.	5	Jet proof: Water projected by a nozzle from any direction (under stated conditions) shall have no harmful effect.
	(b) Dustproof: Protection against harmful deposits of dust, dust may enter but not in any amount sufficient to interfere with satisfactory operation.		
6	(a) Complete protection against contact.	6	Watertight equipment: Protection against conditions on ship's decks, etc. Water from heavy seas or power jets shall not enter the enclosures under prescribed conditions.
	(b) Dust-tight: Protection against ingress of dust.		
IP code notes – Degree of protection is stated in form IPXX. – Protection against contact or ingress of water, respectively, is specified by replacing first or second X by digit number tabled e.g. IP2X defines an enclosure giving protection against finger contact but without any specific protection against ingress of water or liquid.		7	Protection against immersion in water: It shall not be possible for water to enter the enclosure under stated conditions of pressure and time.
		8	Protection against indefinite immersion in water under specified pressure: It shall not be possible for water to enter the enclosure.

N.B. Use this table for general guidance only – refer to BS EN 60529:1992 for full information on degrees of protection offered by enclosures.
* Taken from *The Handbook on the 16th Edition of the IEE Regulation for Electrical Installations* published by ECA, ECAS and NICEIC, 1991, Oxford.

Table 19.2 Typical inspection schedule.

Inspection check list

                                        Area .... Date ....

Conduit
  Fixings correctly applied
  Couplings tightened
  Running coupling locknuts tightened
  Undue sags eliminated
  Box lids fitted complete with all screws
  Vice marks removed
  Damaged finishes restored
  External protective conductors installed

Trunking
  Fixings correctly applied
  Couplers tightened
  Undue sags eliminated
  Lids fitted complete with all screws
  Damaged finishes restored
  Earthing and bonding connections applied

Tray and racking
  Undue sags eliminated
  Couplers tightened
  All fixings correctly applied
  Damaged finishes restored
  Earthing and bonding connections applied
  Cable clips etc. correctly applied
  Cable radii checked

Accessories
  Boxes squarely installed
  Faceplates tightened and complete with all screws
  Damaged faceplates replaced
  Mechanical operation checked

Switchgear
  Fixings secure
  Damaged finishes restored
  Lids tightened and complete with all screws etc.
  Earthing and bonding connections applied
  Mechanical operation checked
  Oil levels checked where applicable
  Dashpots (O/L and E/F) free where applicable
  Relays, packing removed and operation possible

General
  All damage caused by other trades reported

NOTE: Above requirements are additional to Section 712 of the IEE Wiring Regulations.

Appendix 1
Statutory Acts and Regulations

Electricity Supply Regulations 1988.
The Electricity Supply (Amendment) Regulations 1990.
Health & Safety at Work etc. Act 1974.
Electricity at Work Regulations 1989.
The Smoke Detectors Act 1991.
Fire Precautions Act 1971.
The Low Voltage Electrical Equipment (Safety) Regulations 1989.
The Control of Substances Hazardous to Health Regulations 1988.
The Electromagnetic Compatibility Regulations 1992.
Personal Protective Equipment at Work Regulations 1992.
Health and Safety (Display Screen Equipment) Regulations 1992.
Manual Handling Operations Regulations 1992.
Management of Health and Safety at Work Regulations 1992.
Provision and Use of Work Equipment Regulations 1992.
Workplace (Health, Safety and Welfare) Regulations 1992.
Building Standards (Scotland) Regulations with Amendment Regulations 1971–1980.
Cinematograph Acts 1909 and 1952.
Agriculture (Stationary Machinery) Regulations 1959.
The Highly Flammable Liquids and Liquified Petroleum Gases Regulations 1972 (The HF and LPG Regulations).
The Petroleum Act 1928.

Future legislation

The Fire Precautions (Places of Work) Regulations 19–
The Construction (Design and Management) Regulations 19–

Nonstatutory regulations

BS 7671: (1992) Requirements for electrical installations (The IEE Wiring Regulations)

Appendix 2
British Standards and Codes of Practice

Related standards are given in parentheses.
Symbols indicating degree of compatibility:

$=$ A technically equivalent standard

$\neq$ A related but not equivalent standard

$\equiv$ An identical standard.

31: (1988)	Steel conduit and fittings for electrical installations.
88: (1988)	Cartridge fuses for voltages up to and including 1000 V a.c. and 1500 V d.c. ($\equiv$ IEC 269−2).
171: (1978)	Specification for power transformers ($\neq$ IEC 76).
196: (1961)	Specification for protected-type non-reversible plugs, socket-outlets and cable couples.
731: (1980)	Flexible steel conduit and adaptors for the protection of electrical cable.
1361: (1986)	Cartridge fuses for a.c. circuits in domestic and similar premises ($\neq$ IEC 269−1).
1362: (1986)	General purpose fuse-links for domestic and similar purposes (primarily for use in plugs) ($\neq$ IEC 269−1).
1363: (1984)	13 A plugs, switched and unswitched socket-outlets and boxes.
3036: (1986)	Semi-enclosed electric fuses (ratings up to 100 A and 240 V to earth).
3871: Part 1: (1984)	Miniature air-break circuit breaker for a.c. circuits.
3939: (1982)	Graphical symbols for electrical power, telecommunications and electronics diagrams ($\equiv$ IEC 617).
4293: (1983)	Residual current operated circuit-breakers ($\neq$ IEC 755).
4343: (1968)	Specification for industrial plugs, socket-outlets and couplers for a.c. and d.c. supplies ($\neq$ IEC 309−1, IEC 302−2, CEE 17).
4533: (1990)	Luminaires ($\equiv$ EN 60598−1, EN 60598−2−1 $\neq$ IEC 598−1).
4568: (1970)	Steel conduit and fittings with metric threads of ISO form for electrical installations ($\neq$ CEE 23).

4607: (1991)	Non-metallic conduits and fittings for electrical installations ($\neq$ CEE 26).
4678: (1988)	Cable trunking.
4752: (1977)	Switchgear and control gear for voltages up to and including 1000 V a.c. and 1200 V d.c. Superseded by BS EN60947−2.
5266: Part 1: (1988)	Emergency lighting (Code of Practice).
5345: (1989)	Selection, installation and maintenance of electrical apparatus for use in potentially explosive atmospheres (other than mining applications or explosive processing and manufacture) (Code of Practice). ($\neq$ IEC 79, IEC 79−12, IEC 79−14).
5486: Part 1: (1990)	Factory-built assemblies of switchgear and control gear for voltages up to and including 1000 V a.c. and 1200 V d.c. ($\equiv$ EN 60439−1 $\neq$ IEC 439−1).
5501: (1977)	Electrical apparatus for potentially explosive atmospheres (Euro-standards $\equiv$ EN 50014/20 $\neq$ IEC 79−0).
5839: Part 1: (1988)	Fire detection and alarm systems in buildings (Code of Practice).
6004: (1991)	PVC-insulated cables (non-armoured) for electric power and lighting ($\neq$ IEC 227).
6081: (1989)	Terminations for mineral-insulated cables ($\neq$ IEC 702−2).
6099: (1986)	Conduits for electrical installations ($\equiv$ IEC 614).
6207: (1991)	Mineral-insulated cables ($\neq$ IEC 702−1: 1988).
6651: (1992)	Protection of structures against lightning (Code of Practice).
7430: (1991)	Earthing (Code of Practice).
7671: (1992)	Requirements for electrical installations (The IEE Wiring Regulations).

Codes of Practice (earlier editions)

1003: (1967)	Electrical apparatus and associated equipment for use in explosive atmospheres of gas or vapour other than mining applications (Replaced by Parts 1 to 8 of BS 5345 but retained as a reference guide for existing installations)
1007: (1955)	Maintained lighting for cinemas.

1019: (1972) Installation and servicing of electrical fire alarm systems (withdrawn and superseded by BS 5839: Part 1: 1980).

BSI address

British Standards Institution
Sales Department
389 Chiswick High Road
London
W4 4AL

Index